Florencia Aspera
Eugenia Artola

Documentary review of the Pythagorean Theorem throughout history

Florencia Aspera
Eugenia Artola

Documentary review of the Pythagorean Theorem throughout history

Didactic sequences for classroom work with middle school students

ScienciaScripts

Imprint

Any brand names and product names mentioned in this book are subject to trademark, brand or patent protection and are trademarks or registered trademarks of their respective holders. The use of brand names, product names, common names, trade names, product descriptions etc. even without a particular marking in this work is in no way to be construed to mean that such names may be regarded as unrestricted in respect of trademark and brand protection legislation and could thus be used by anyone.

Cover image: www.ingimage.com

This book is a translation from the original published under ISBN 978-620-2-12578-9.

Publisher:
Sciencia Scripts
is a trademark of
Dodo Books Indian Ocean Ltd. and OmniScriptum S.R.L publishing group

120 High Road, East Finchley, London, N2 9ED, United Kingdom
Str. Armeneasca 28/1, office 1, Chisinau MD-2012, Republic of Moldova, Europe
Printed at: see last page
ISBN: 978-620-6-97199-3

Contents

To my children Martina, Josefina and Bernardo, whom I love deeply.

To my husband, my great partner.
To my mother, the strongest woman I know.
To my siblings Cristian, Andrea, Carolina and Sebastian, with whom I walk with in this life.

ACKNOWLEDGEMENTS

I thank my Director Eugenia Artola for the love with which she accompanied me on this journey, for dedicating her time and patience in reviewing, correcting and suggesting improvements to this document. I would also like to thank Ariana Dalvelo for her suggestions and advice. They are references for me in my profession and in life.

I thank my children for their unconditional love. To my husband whom I love, who accompanied me in the long nights of writing with coffee and patience. To my mother and my siblings. To my niece Emma and my brother-in-law Santiago.

To my History of Mathematics teacher Mirta Di Cësare who passed on to me her love for mathematics and its history. And to the teachers I had during all my school and university years.

To my friends and colleagues Gabriela, Pamela, Rita and Martin, who studied with me the Bachelor's Degree in Mathematics, who always encouraged me to go ahead.

SUMMARY

The Pythagoras theorem, as Kepler (1596) says, is one of the greatest treasures of Geometry, it states that *"in every right triangle the measure of the hypotenuse squared is equal to the sum of the measures of the squares of the legs"*. Anyone who went to secondary school has come into contact with this theorem, whether they remember it or not.

The general objective of this work is to analyse the proofs and demonstrations of the Pythagorean Theorem throughout the History of Mathematics, with the aim of creating didactic sequences for its teaching at the intermediate level.

It is organised in five chapters, the first of which presents the problem statement, objectives and justification.

In Chapter 2, referring to the theoretical framework, emphasis is placed on the socio-epistemological vision of the term mathematical "demonstration", as a means of proof and as a mathematical object in itself, differentiating the terms "demonstration", "proof" and "explanation", which have repeatedly been used as synonyms in educational practices. A taxonomy of proofs is then proposed.

Chapter 3 presents proofs and demonstrations of the Pythagorean Theorem throughout history, and the chapter ends with a classification of them according to the taxonomy displayed in the previous chapter.

In Chapter 4, two didactic sequences are presented, starting by analysing mathematical demonstration from a didactic point of view, then describing the tasks to address geometric notions, and focusing on demonstration tasks, as well as considering the importance of incorporating ICT into the proposals. It culminates with the didactic sequences that incorporate the previously mentioned, and that aim to awaken the interbs and creativity in the design of didactic sequences for the Middle Level related to Geometry, and mainly to the Pythagorean Theorem based on proofs carried out throughout the History of Mathematics.

Chapter 5 presents the results of the survey carried out prior to the redaction of this work, and the conclusions of this work.

Chapter 1

Problem statement

"A great discovery solves a great problem, but in the solution of every problem, there is a certain discovery. The problem posed may be modest, but, if it tests the curiosity that brings the inventive faculties into play, if it is solved by one's own means, one can experience the charm of discovery and the joy of triumph".

George Polya (1984)

1.1 Introduction

Mathematics is a formal science that is responsible, through axioms and properties, for showing relationships between mathematical objects. It differs from other areas of knowledge, among other aspects, because its study is of a semiotic nature, i.e. mathematical notions are studied through the representation of their objects (Damisa and Ponzetti, 2015, p.136).

The Pythagorean Theorem is a relation that is verified in right-angled triangles and states that: *"the measure of the square of the hypotenuse is equal to the sum of the measures of the squares of the legs"*. Through graphical and algebraic representation, throughout history ancient civilisations, mathematicians and others have proved and demonstrated this eorem, giving rise to a great variety of proofs. Inquiring about them will allow us to make them known and use them to carry out didactic sequences that enrich teaching practices at the intermediate level. In this research we will analyse various proofs and demonstrations of the Pythagorean Theorem throughout the History of Mathematics.

The problem of investigation lies in the loss of prominence of the teaching and learning of Geometry in the planning and programmes of Secondary Schools. Within this branch of Mathematics, the Pythagorean Theorem is one of the knowledge with the highest level of application in intramathematical contexts such as Geometry, Calculus, Trigonometry and Algebra, and extramathematical contexts such as Cartography, Navigation, Engineering, Architecture, among others. Beyond the importance of the Pythagorean Theorem in Mathematics and in other areas, its teaching is also re-signified in the teaching of demonstrations and verifications at the intermediate level. The development of this skill will allow students to reach new levels of understanding.

But why is it important for middle level students to know and work on some of the verifications of the Pythagorean Theorem in the classroom? For Hanna and Jahnke (1996; p. 887) "demonstration is an essential feature of mathematics and as such should be a key component of mathematics education". The advantages of demonstration work in the classroom will lead to the development of students' skills in manipulating the mathematical object, justifying points of view, validating and communicating results, among others.

In this research we will analyse various proofs and demonstrations of the Pythagorean Theorem throughout the History of Mathematics, considering the following aspects:

- The underlying importance of teaching the Pythagorean Theorem at the intermediate level,
- The contribution to the teaching of evidence throughout history,
- The theoretical and didactic power of this Theorem,
- Interests and creativity in the design of didactic sequences for the intermediate level.

1.2 Survey conducted

Before starting this work, 79 people were surveyed, 24% of whom were teachers. The purpose of the survey was, for the teachers, to find out about the method of teaching the Pythagorean Theorem, and for the rest of the respondents, to find out about their knowledge of the Pythagorean Theorem, i.e. whether they remembered it, whether they could state it, whether they proved or demonstrated it, and whether they applied it later in their lives.

The survey was conducted using a Google Drive Questionnaire, and was shared via email, whatsapp and Facebook (see Annex 1 for the completed survey).

The data obtained from the questionnaire were as follows:

- 24% of the respondents were mathematics teachers.
- Of these teachers, 84% agreed that the Pythagorean Theorem is taught between the first and second year of secondary school, 37% for the first year and 47% for the second year.
- When asked to state the theorem, 37% stated it completely and correctly. The rest of the respondents mentioned the statement as the formula that allows to obtain one of the unknown sides of the theorem or to corroborate whether a triangle is rectangular or not.
- While 95% of the teachers responded that they do carry out verifications of this curricular content when they teach it, 84% do so by means of geometric tests: based on comparisons of areas, and 11% with algebraic tests. When they refer to geometric proofs, they do so by comparing areas of squares on each side. The latter is one of the many ways in which these verifications can be carried out, and throughout this work we want to show that there are other verifications of the theorem, since, as Duval (1993) states, to the extent that students manipulate with different representations of the same mathematical object, this has a direct impact on the acquisition of the concept.
- As for the moment in which the verification of the Theorem is carried out, 47% do it before teaching it formally, 37% after stating it and the remaining 16% as an exercise.
- All teachers consider it to be very relevant content, justifying it by the variety of applications it has.

The survey and its results, promoted the writing of this work and the need to make a reservoir of proofs and demonstrations of the Pythagorean Theorem, of which teachers can avail themselves for their work in the classroom. In addition, to guide and encourage their own proposals, so as to allow a better process of teaching and learning of the discipline in the middle level of the schools of the Province of Mendoza.

1.3 Objectives

General Objective:

- To analyse the proofs and demonstrations of the Pythagorean Theorem throughout the History of Mathematics, in order to create didactic sequences for its teaching at the intermediate level.

Specific Objectives:

- To present proofs and demonstrations of the Pythagorean Theorem throughout the History of Mathematics.
- Design didactic sequences related to the proofs of the Pythagorean Theorem for the intermediate level.

Transfer Objective:

- To awaken the interest and creativity in the design of didactic sequences for the Middle Level related to Geometry, and mainly to the Pythagorean Theorem based on proofs carried out throughout the History of this science.

1.4 **Justification**

The Pythagorean Theorem is a mathematical relation present in the Secondary School Mathematics syllabus. For some time now, there has been a great concern in secondary school teaching about the question of why geometry is less important in the planning and syllabuses at this level of education. This leads to a reflection on why geometric notions should be taught in secondary school. In the article entitled "Espmtu de geometria" (EL PA^S, 5/12/99), which belongs to the series "El dardo en la palabra", F. Lazaro Carreter begins with this sentence: ^Podriamos hablar sin la Geometria? It slips through all the seams of the language, without us even realising it...

According to Santalo (1975), Mathematics has always been part of the whole educational system, and going back to the Hellenistic world he emphasises that in ancient Greece the first pillars of Education were Arithmetic and Geometry, as described by Plato in The Republic. There, Socrates and Glaucon established that Geometry corresponded to the second teaching of the time, after Philosophy. Geometry was esteemed because it was regarded as a body of knowledge that was true, indeed it could be proved to be true.

In ancient times, geometric knowledge was acquired from intuition, however, in schooling, students have increasingly acquired a deductive method, turning to the memorisation of concepts, theorems, properties and formulas. Alsina, Burguds and Fortuny (1987) point out that the distinction between what is *useful to learn* and what is *desirable to teach* was a current issue. This paper considers that this gap, from 1987 to the present day, has widened and may be one of the causes that lead to the loss of prominence of Geometry teaching at the intermediate level. Young people are prepared for the future by teaching them only what is believed to be useful for their development in society, but without paying attention to the fact that changes in society are taking place more rapidly than those occurring at the educational level. Thus:

What is extraordinarily useless is that which is unsuitable for the level and ability of the learner. For example, constructions with rulers and compasses are useful and formative, but they are useless at an age when these instruments cannot be handled manually and with ease or when the definition of straight line and circumference is not assumed ... *what is desirable in the teaching of geometry is that which is useful in the future and can be motivated from the present time*: reasoning correctly (deductively and inductively), representing, abstracting, relating, classifying and solving are key verbs in the range of what is desirable (Alsina, et al., 1987, p. 18), 1987, p. 18).

Kline (1978) states:

It is intellectually dishonest to teach deductive interpretation as if one arrives at results by pure logic [...] Such interpretation destroys the life and spirit of mathematics [...], and although it has an aesthetic appeal for the mathematician it serves as an anaesthetic for the student (Kline, 1978, pp. 54-60).

In order to reduce this great gap, the History of Mathematics can be considered as an excellent didactic resource, which brings together what is useful and what is desirable to teach, but bearing in mind that for this the teacher must be interested in adapting, reconstructing, recreating and transforming the institutionalised historical knowledge into the knowledge to be taught, thus carrying out a *didactic transposition* as Chevallard (1985) points out.

In this way, the teacher can transmit his knowledge by de-dramatising the Teaching of Mathematics, considering that "the History of Mathematics is an inexhaustible source of didactic material, of interesting ideas and problems and also, to a great extent, of amusement and intellectual recreation, in short, of personal, scientific and professional enrichment"

(Gonzalez Urbaneja, 2004, p. 27).

Gonzalez Urbaneja (2004) points out that the historical use of mathematics can be done in different ways:

• Precede the exposition of each topic with a historical introduction, placing the origin and evolution of the problems to be dealt with in the scientific and cultural contexts.

• Classroom records can be supplemented with prompts, brief summaries or historical notes.

• Throughout the course of the lesson and at any time, it is possible to briefly indicate to which mathematician or mathematical current the introduction of a new concept, the proof of a theorem or the solution of a problem is due (p.22).

In this work, the Pythagorean Theorem and its historical evolution are used as a didactic resource, because, as indicated by (Gonzalez Urbareja, 2001a) cited by Gonzalez Urbareja (2004):

The Pythagorean Theorem (whose appearance on the mathematical historical horizon, but also on the school horizon, marks the first intellectual leap between the confines of empirical speculation and the domains of deductive reasoning, since with it comes the historical and school phenomenon of demonstration, so that it is a real paradigm for mathematics and above all for mathematics education (p. 22).

Thus, in this paper we will use the terms *proof* and *demonstration* as defined by Balacheff (2000), according to the context in which they will be applied for the analysis of the Pythagorean Theorem throughout history. "A context or institutional framework is considered to be a local point of view or perspective on a given problem, characterised by the use of its own expressive and instrumental resources, by specific habits and norms of behaviour" (Godino and Recio, 2001, p. 407). Therefore, the institutional framework that will be taken for the demonstration in this work is the Mathematics class and will be developed in Chapter 4.

In section 2.2. a taxonomy of proofs will be made, to be applied to the proofs and demonstrations of the Pythagorean Theorem made throughout history, which are mentioned in Chapter 3.

In this research, access to the historical construction of knowledge, in this case the Pythagorean Theorem, is considered fundamental, taking as a basis the historical proofs and demonstrations of this theorem, in order to elaborate didactic sequences that allow the basic competences of mathematical work to be deployed.

Chapter 2

Theoretical Framework

"The fundamental and general task of the mathematical community is to contribute effectively to the integral development of human culture".

(de Guzman, 1996)

2. 1 Demonstrations in Mathematics from a vision socio-epistemology

The concept of mathematical proof has evolved notably throughout history (Arsac, 1987). The idea of what a demonstration is and when it is considered valid is relative to the socio-cultural scenario in which we place ourselves and varies considerably from one culture to another (Crespo Crespo, 2005, p. 16). Several authors consider that the origin of demonstration dates back to the 5th century BC in ancient Greece. But before the Greeks there were no demonstrations? The answer is no, because the term would be anachronistic. Before Pythagoras, observation and measurement, in some cases, were part of the geometrical ideas of the time and were considered as means of proof. Heuristically, a proof is a rhetorical device to convince another person that a mathematical statement is true or valid (Krantz, 2007, p.3). As will be discussed below, perhaps the first recorded mathematical "proof" in history is attributed to the Babylonians.

Establishing the etymological origin of the word demonstration goes back to Latin, as it derives from *"demonstratio"*. It is a term that is composed of several Latin elements: the prefix *"de"*, which is used to indicate a separation, the verb *"monstrare"*, which can be translated as to show, and the suffix *"cion"*, which is used to indicate an action and its effect. Indicating, pointing out, showing or proving something involves an action known as demonstrating. This activity and its effects are called demonstration.

One of the aims of this work is to analyse the proofs and demonstrations of the Pythagorean Theorem throughout history. As mentioned, before the Greek civilisation, the study of mathematical notions was based on an inductive method, where from several repeated observations a knowledge could be induced. However, we should not underestimate the importance of the mathematics developed in Babylon, Ancient Egypt, Ancient China or Ancient India, among other civilisations, which have managed to carry out surprising proofs in relation to their time, using everyday problems or situations which they modified to show their often approximate results. This does not mean that in the proofs they carried out, the results of the calculations used have always been approximate, a clear example of this fact is the development of the Pythagorean triads and the Pythagorean Theorem; despite the fact that many authors have made a distinction between Ancient Eastern and Ancient Greek Mathematics, arguing that the former is approximate, since they used approximations in their calculations; and the latter is exact, because it works in an abstract way and not with concrete numbers.

As mentioned, Greek mathematicians used deductive reasoning, they used logical reasoning in such a way that from definitions and axioms, conclusions, properties or theorems could be deduced. This idea of mathematics as a framework of theorems supported by axioms is expounded in Greek texts referring to geometrical notions. In context, Geometry (from the Greek: ysropsrpia: geo (Earth), metria (measure), according to the Royal Spanish Academy, studies the properties and magnitudes of figures in the plane or in space. From its beginnings,

Geometry was concerned with providing answers to everyday problems and the development of classical Geometry, which is based on Euclid's Elements (around 300 BC), is defined as the science of geometrical figures. It was the first branch of mathematics to consolidate the mathematical knowledge of its time, to organize and formalize it. Nowadays this branch of mathematics has been influenced by algebra and calculus, giving rise to modern geometry. A clear example of this change is the following:

Another classical problem consisted in proving the Fifth Postulate of Euclid's Elements ("from a point outside a straight line one can only draw a parallel to it") from the remaining Euclidean postulates. For centuries there were unsuccessful attempts. The problem reached its true dimension through its inversion. Instead of attempting the direct proof, which presupposes the admission of the unique parallelism, it is denied in various ways (there are no parallels - elliptic geometry -, or, there are infinite parallels - hyperbolic geometry -). Thus, several logically correct non-Euclidean geometries (Gauss, Lobachetvski, Bolyai, Riemann) were simultaneously achieved and, as it was later proved, also physically feasible. This opened up a crisis in the foundations of mathematics: from this point on, the relationship between mathematics and reality was posed from new perspectives, as well as the very concept of mathematical space, which now became questionable (Sorando, n.d.).

In this paper we consider Euclidean Geometry, as enunciated by Hilbert:

Elementary (Euclidean) geometry is concerned with the facts and laws that the spatial behaviour of things presents to us. According to its structure, it is a system of propositions which - to a greater or lesser extent - can be deduced in a purely logical way from certain undemonstrable propositions, the axioms. This behaviour, which in lesser completeness we find, for example, in mathematical physics, can be expressed briefly in the sentence: Geometry is the most complete natural science (Hilbert, 1898/1899, p. 302).

The Pythagorean Theorem was already known in civilisations prior to the Greek, however, we cannot take away the merit of Pythagoras of Samos, born between 582 BC and 569 BC, since he managed to extend its veracity to all right-angled triangles, and not to particular cases already known. The statement that appears as proposition 47 of the first book of Euclid, translated by Clark University states that:

In right triangles the square of the side opposite the right angle is equal to the sum of the squares of the sides comprising the right angle.

It has now been amended as follows:

In any right triangle, the area of the square whose side is the hypotenuse is equal to the sum of the areas of the squares whose sides are each of the legs.

Moreover, Pythagoras was one of the first mathematicians who created a closed group of disciples who participated in his studies and defended his doctrine. In this way a new mathematical universe emerged, with a mystical vision of numbers and seeking their relationship with geometrical notions. Geometry was not always a collection of statements, axioms and demonstrations. In its origins this branch of mathematics was based on a collection of empirically discovered propositions relating to lengths, angles, areas, and volumes of various objects, which were developed to satisfy needs in surveying, construction, astronomy, and craftsmanship.

In the field of mathematics, it is essential to bear in mind that the logical sequence of propositions, starting from axioms or postulates that are used to arrive at the theorem to be proved, cannot be taught in the same way in a classroom as in a purely scientific environment. In order to become teaching content, mathematical notions must undergo an adaptive transformation, a didactic transposition (Chevallard, 1985). Moreover, it must be taken into account that the new curricular orientations demand the development of mathematical reasoning skills and in particular the ability to carry out mathematical demonstrations. Prior Martinez and Torregrosa Gironds (2013) state that it is a complex matter to achieve this goal, since several studies conclude that students do not feel the need to perform deductive demonstrations (Hanna and Jahnke, 1996), and do not distinguish between the different forms of mathematical reasoning involving explanation, argumentation, verification and demonstration (Dreyfus, 1999).

Characterising the development of cognitive processes involves the study of the interaction between reasoning processes and the verification procedures used in the solution of geometric problems, and in particular in the Pythagorean Theorem. Consequently, the use of different verification procedures to establish the truth of a proposition is related to the different outcomes of reasoning in problems that require a proof (Prior Martinez and Torregrosa Gironds, 2013).

In other fields of knowledge, the truth of a proposition is obtained from data coming from perception, measurements, testimonies, etc.; however, the truth of a mathematical proposition is obtained deductively from a succession of other previous propositions that are true (Duval, 2007). Although in these deductions the idea is to justify or validate a statement (thesis) by providing reasons or arguments, differences can be observed in the situations in which they are used, in the characteristic features and in the expressive resources brought into play in each of them. The differences found in the situations and argumentative practices indicate different senses of the concept of demonstration (Godino and Recio, 2001).

For Balacheff (1987) the term "explication" is primitive and implies the terms "proof" and "demonstration". The purpose of explication is to establish in the interlocutor a system of objects characterised by a certain homogeneity. These objects meet, harmonise and in their affinity determine the organisation of an explanation which is oriented towards the discovery of a new knowledge (Midville, 1981, p. 150). However, Duval (1993, p.40) considers that in "argumentation" it is a matter of showing the truth of a proposition, whereas in "explanation" the statements have a descriptive intention of a phenomenon, result or behaviour. In this way, Balacheff's notion of "explanation" is assimilated to Duval's notion of "argumentation" (Godino and Recio, 2001).

In the mathematical community, only those explanations that take a particular form as sequences of statements organised according to certain rules, which are deduced from others that precede them, are accepted as "proofs" (Balacheff, 1988). Therefore, for an explanation to be a proof it has to be seen or recognised by someone as sufficient reason in the corresponding discursive framework (Godino and Recio, 2001). The researchers Harel and Sowder (1998) studied students' conceptions of propositional validation and proposed a classification of proof schemas which they defined as their methods of verifying mathematical claims. They classified these proof schemes into external, entrepreneurial and anathematic conviction schemes. A *verification procedure* is defined as the proof scheme used by the student to verify, from his or her point of view, a mathematical statement that is part of the solution to a given geographical problem, or to establish the truth of this or that statement. In other words, it is the procedure that is carried out to establish the truth of a proposition, which is used in the chain of reasoning to demonstrate a certain geomorphic property requested (Prior Martinez and Torregrosa Gironbs, 2013).

These procedures can be classified as follows:

• *Perceptual verification procedure*: beyond the register of language, there is an access to what is represented given by the content of the propositions. This access can be immediate or instrumental, i.e. direct or subject to experimental procedures.

• *External verification procedure*: the certainty of the proposition comes from the fact that others agree with its truth (teacher, rest of the group, prior knowledge, etc.). The statement is taken as true and this value is not questioned.

• *Deductive verification procedure*: the proposition can be deductively placed in a series of other propositions, where the previous propositions have truth value.

In school practice, verification procedures are different from those used by mathematicians, since the latter, when they wish to prove a proposition, make use only of previously proved propositions, such as theorems or axioms of the theory in which they are situated. On the other hand, intermediate level students often accept certain propositions as valid by means of external verification procedures, and consequently, when a student associates a mathematical statement with a representation that forms part of his or her school experience, this association is valid, even if the student cannot prove its veracity (Prior Martinez and Torregrosa Gironbs, 2013).

For Godino and Recio (2001, p. 406) reasoning "is an intellectual activity, which cannot be reduced merely to the manipulation of information, but which gives rise to argumentative practices, personal or institutional, which constitute its ostensive and communicational dimension. At the same time, reasoning develops through these practices, so that the study of reasoning is constitutively linked to the study of argumentation. Balacheff (2000) addresses reasoning in the study of the process of proof and validation situations, stating that reasoning is an intellectual activity that attempts to modify and produce new considerations. He mentions a social dimension of proof by involving the community that accepts or does not accept reasoning, and considers that it is necessary to recall the two main aspects of demonstration as they relate to the scientific community:

• *Demonstration is an essential testing tool; it leads to a practical exercise, which makes communication and evaluation possible at the same time;*

• *It is also the object of study of those who deal with logic; demonstration has, therefore, a precise definition within the framework of formalised theories.*

The term "demonstration" has similar meanings for Balacheff and for Duval (cited in Godino and Recio, 2001), who define it as "a sequence of statements organised according to given rules", since a statement is conventionally assumed to be true, or else it is deduced from those preceding it with the help of rules of deduction taken from a well-defined set of rules whose validity is socially shared.

The Greeks valued geometry as a body of true knowledge that can be demonstrated, which is why in Plato's school of philosophy it was written: "*Let no one enter here who does not know geometry*". *In* other words, Geometry is not valued today as it was by the Greeks and, as Balacheff (2000) points out: "Mathematics teaching unconsciously strips students of the responsibility for truth". Lectures are given in an informative way, without questioning, ostensibly showing figures, verifying properties and simply solving exercises that are reduced to the mere application of formulas. Brousseau (1986, p. 359) states that "didactics is faced with the challenge of producing situations that allow the student to put mathematical knowledge and skills into practice as an effective means of convincing (and being convinced) of everything, leading him to reject rhetorical means, which are neither good proofs nor refutations".

The verbs explain, prove and demonstrate are often considered as synonyms in the practice of teaching mathematical notions, however, Balacheff (2000) considers that the vocabulary used in mathematics classes must be precise, since the result obtained from the students will depend on this. To this end, he defines them as follows:

- **Explanation:** Following the linguists, we place explanation at the level of the speaker subject. For him it establishes and guarantees the validity of a proposition, it is rooted in his knowledge and in what constitutes his rationality, i.e. his own rules for deciding the truth.
- **Proof:** means a justification where the validity can be modified depending on the experience of the community and time. The passage from explanation to proof is given by the acceptance of a community, of course there may be another community that does not accept it.
- **Demonstration:** means a proof that takes a particular and logical form, based on a strongly institutionalised body of knowledge, i.e. a set of definitions, theorems and rules of deduction whose validity is shared by the mathematical scientific community. It is a series of statements that are organised according to a well-defined set of rules. What characterises demonstrations as a genre of discourse is their strictly codified form (Balacheff, 2000, p.13).

The conception of mathematical rigour is an implicit one that depends on the school institution, and its appropriation by students will require a special cognitive construction, which is not spontaneous and should be taught; moreover, a proof is an explanation accepted by a given community at a given time, and a demonstration is a particular case of these proofs under the rules of mathematical deduction that are given within the scientific community (Balacheff, 1987).

The so-called formalist current emerged towards the end of the 19th and beginning of the 20th century, was promoted by Hilbert and conceives mathematics as the result of the manipulation of symbols based on certain rules. According to Crespo, Farfan and Lezama (2010), it is a perspective proper to logic and under this paradigm, the term demonstration is defined as a sequence of propositions, each of which is either an axiom or a proposition that has been derived from the initial axioms by the rules of inference of the theory. The difference that Crespo et al. (2010) point out is that, if only axioms and rules of inference are used, developing the derivation in a complete way, it is said to be a *demonstration*. But if use is made of previously proved theorems, it is said to be a *proof*.

Garda and Lopez (2008) also propose definitions for explanation, proof and demonstration, which are developed in the school environment.

- **Explanation:** is a discourse that tries to make intelligible the truth of a proposition or a result. The reasons given can be disputed, refuted or accepted.
- **Proof:** is an explanation accepted by a given community at a given time, it can be the subject of a debate whose significance is to determine a common system of validation among those involved in the discussion of proof.
- **Demonstration:** in the school classroom what is really done is to prove that certain statements are true; they are not considered rigorous demonstrations because they are not part of an axiomatic system based on definitions and axioms.

While Duval (1999) defines the following terms:

- **Explanation:** is to give one or more reasons to make a data, phenomenon or result understandable.
- **Demonstration:** a sequence of statements organised according to given rules.

For Balacheff (2000) an explanation is a discourse that is related to the set of knowledge that is developed in the discovery of a new knowledge and that is put into play in front of an interlocutor. It does not necessarily have to follow deductive steps, but tries to argue from one's own perspective the veracity of something.

For the authors Garda and Lopez (2008), an explanation is also a discourse to demonstrate the veracity of a statement, which can be accepted, disputed or refuted by the interlocutors. For these authors, an explanation does not have scientific rigour as it is not based on an axiomatic system. As for proof, they consider it to be an explanation that is accepted by a scientific community even if it does not have sufficient scientific rigour. Finally, when explanations and proofs are based on an axiomatic system, they are considered as a demonstration. However, the authors agree that in the classroom it is often not possible to acquire the formal level of a demonstration, since students lack knowledge about the axiomatic system. It is worth noting that both authors Garcia and Lopez (2008) agree with Balacheff (2000) in the definitions mentioned. For Godino and Recio (2001), demonstration is the emerging object of the system of argumentative practices (or arguments) accepted within a community, or by a person, in situations of validation and decision, that is, situations that require justifying or validating the truth of a statement, its consistency or the effectiveness of an action (p. 406).

Table 1 is a comparative table of the different definitions that have been shown for the terms *demonstration, proof, explanation* and *argumentation*, according to different authors, and is therefore a synthesis of what has been analysed in this section.

Table 1. Comparative table of the terms demonstration, proof, explanation and argumentation, according to different authors.

Authors	*Demonstration*	*TestExplication*
	The deductive process led to The deductive process carried out in order to obtain a out to obtain a theorem.　　o　　　11T1 11 TO　　　п 1A 1 Γ'o m /ATA TZA theorem uses only	
Vision Formalist of Hilbert	uses only axioms and axioms and rules of inference, 　　　rulesfrom　　inference, in the development of drift developing the derivation of is made use of theorems comprehensive manner. previously demonstrated.	
Balacheff	It is a discourse that this particular form, given its validity, can be modified in relation to the whole of the 　　validity may be modifiable to related to the set of The discovery of a new knowledge, which is developed by a series of rules depending on the experience of the community and the time.　discovery of a new knowledge and which are brought into play in front of an interlocutor.	
Duval	It is a sequence of　statementsIt is to give one or more reasons for organised according to　　　to make understandable a determined rules.　　data, phenomenon or result.	
Godino and Recio	It is the emergent object of　theIt is the discourse that aims to system of *practices* to make intelligible the character of *argumentative* (or arguments)truth, acquired for the accepted within an author　　, a proposition or a community, or by a person,　an outcome. in situations of validation and decision, i.e. situations that require justifying or validating the truth of a statement, its consistency or the effectiveness of an action.	
Garcia and Lopez	In the school classroom, what is really being done is to prove that a given community at a given time is an explanation accepted by a given community is a discourse which tries to make certain statements true; they are not considered to be the subject of a debate whose result. The reasons given are rigorous demonstrations 　　demonstrations　　　meaning is to determine a can be disputed, refuted or refuted. because they are not part of a common or accepted system of validation. axiomatic system that starts from among those involved in the definitions and axioms.　　discussion of the proof.	

Florence Veronica Aspera

33

2.2 Types of Tests

2.2.1 Pragmatic and Intellectual Tests

According to Balacheff (1987), explanations emerge from reasoning that are accepted or not by a specific community; those explanations that are accepted make up the evidence, of which pragmatic and intellectual evidence are distinguished.

He proposes the following classification of evidence (Balacheff, 2000, p.22):

• **Pragmatic proofs***:* are those that resort to action on objects and presuppose the possibility of having access to the material performance of a task in order to justify affirmations about them. They are linked to action and experience, there is a presence of practical knowledge, and justifications are made through concrete material or representation of the object.

• **Intellectual proofs***:* these are those which, separating themselves from action, are based on formulations of the properties at stake and their relations. They come from a particular way of reasoning, where arguments, chains of arguments, are articulated, with a clear production in a symbolic language, there is a passage to the algebraic, material objects and their relation to material experience are left aside.

The evolution of the language and the knowledge of the person who carries out a pragmatic proof constitute the essential characteristics to be able to make the passage towards an intellectual proof, more precisely to a demonstration. The use of an adequate and functional language requires the decontextualisation of the objects of study, which allows a generality and also a depersonalisation and detemporalisation so that the demonstration is perennial.

2.2.1.1 *Practical tests, verification tests, tests*
by means of a diagram or generic model, visually supported evidence and formal
demonstration

Puig and Cerdan (1995) propose a different and complementary classification for the tests, these authors follow the research of Guilldn (1997) and Vega (1990), considering the following types of tests:

• **Practical evidence:** shows, proves or exhibits with a device to establish the truth.

• **Verification tests**: consist of testing or verifying experimentally with one or more examples.

• **Proof by means of a diagram or generic model**: may be incomplete, but is accompanied by some properties of the geometric object and may include some possible generalisation.

• **Visually supported tests**: informal deductive reasoning based on the visual aspect is used, the language is not perfectly formalised and the connection of concepts used is quite simple.

• **Formal demonstration**: proofs accepted in the scientific community, as an organisation of a sequence of formal implications, based on the hypotheses of the problem and on other elements of the axiomatic system such as definitions, properties, theorems.

There is an affinity between the classification given by Balacheff (2000) and these authors, since *practical* proofs, proofs of *verification* and proofs *by means of a diagram* would constitute *pragmatic* proofs, i.e. those proofs related to naive empiricism, crucial experience and generic example. On the other hand, proofs *with visual support* and *formal demonstration* constitute *intellectual* proofs, since they are based on formulations of the properties in play and their relationships.

2.2.1.2 *Algebraic proofs, geometric proofs, dynamic proofs and quaternionic proofs*

Loomis (1968) exhaustively compiled multiple proofs for the Pythagorean Theorem throughout history and further classified the proofs into four types:

• **Algebraic tests**: based on relations between sides and segments.

• **Geometric tests**: based on area comparisons.

• **Dynamic tests:** based on the concepts of mass, speed, strength, etc.

• **Quaternionic tests**: based on vector operations.

From what has been seen above, this research work suggests a taxonomy of proofs, considering verifications, explanations, demonstrations and the different types of proofs according to Balacheff (2000), Loomis (1968) and Puig and Cerdan (1995), which will be applied in the analysis of the proofs and demonstrations that have been carried out throughout history for the Pythagorean Theorem:

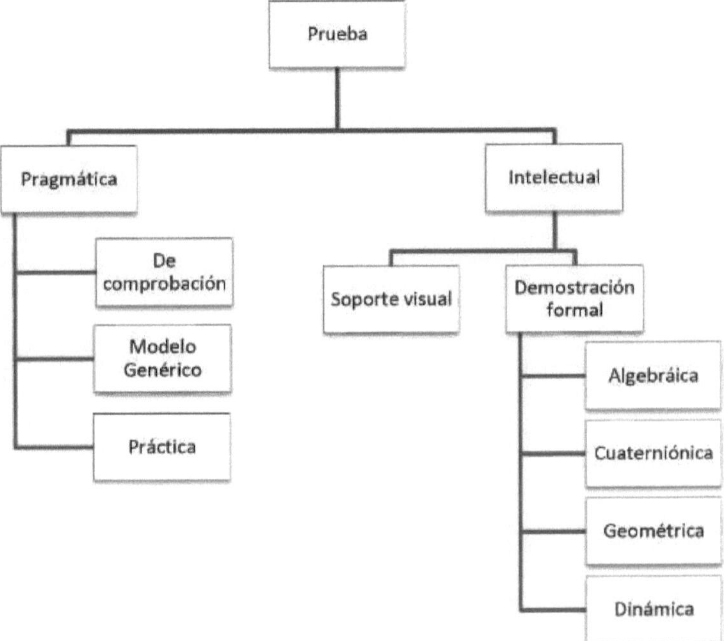

Figure 1. Taxonomy of Evidence

Chapter 3

Historical proofs and demonstrations of the Pythagorean Theorem

The Pythagorean Theorem has kept mathematicians busy from classical times to the present day. [...] Its multiple proofs illustrate the agility of mathematicians in attacking the same problem from different angles. [...] No matter how often it is proved, the Pythagorean Theorem always manages to retain its beauty, its freshness and its eternal sense of wonder.

W. Dunham, 1995. Chap.H. pp.136, 153.

3.1 The Pythagorean Theorem in Babylonia

Cuneiform writing was used for a period of approximately 3400 years, from 3300 BC to 100 AD (Seri, 2015), and developed in the city of Uruk (Mesopotamia) in what is now the Republic of Iraq. They were recorded on clay tablets that over the years were discovered, one of them being the tablet YBC 7289 dated in Paleobabylonian times (2000-1600 BC) and shows that the Babylonians already knew the famous Pythagorean Theorem before the Greek genius described it (Tostado, 2013). How was this possible?

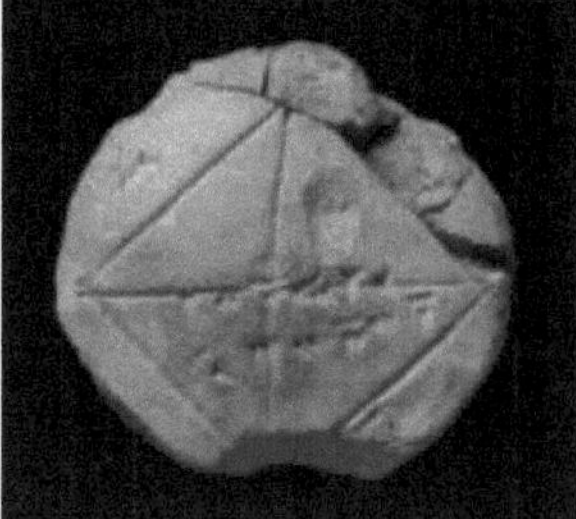

Figure 2. Tablet YBC 7289

Source: https://babylonian-collection.yale.edu/highlights

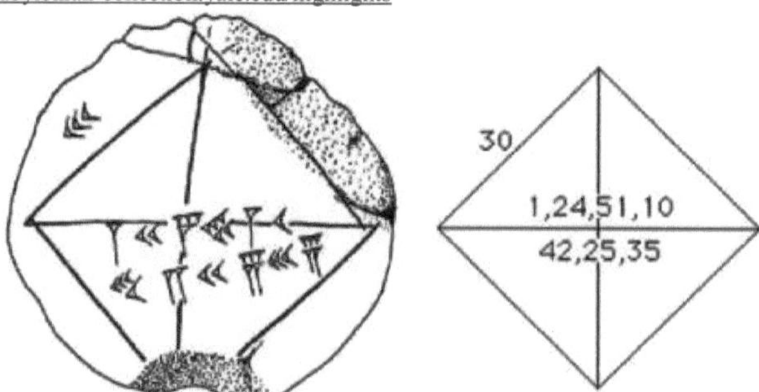

Figure 3. Explanatory Drawing of YCB 7289

Source: https://ipch.yale.edu/news-events/3d-print-ancient-history-one-most-famous-mathematical-texts-mesopotamia

Figure 2 is a photograph of the original YCB 7289 tablet, while figure 3 is an explanatory drawing of the original YCB 7289 tablet.

an explanatory drawing of the tablet can be seen in figure 3. Both show a square with cradle

18

marks (hence the name cuneiform).

The square is a square with cradle marks (hence the name cuneiform) representing numbers. The square

has 30 mm of side, with both diagonals marked, at the same time in the horizontal one we can find

numbers in the form of sexagesimal numeration (a positional numbering system that uses the sexagesimal as its arithmetical base).

arithmetical basis of the number 60). A sexagesimal system is one in which the non-integer magnitudes are found by dividing the number by 60, and is widely used in the measurement of angles and time. According to Boyer (quoted in Martinez, 1986, p.50) the Babylonians established that, by grouping the same symbol in different ways, they could determine the relative position unambiguously when reading from right to left, this corresponds to the successive increasing powers of the base. Figure 4 below shows the Babylonian numbering system.

Figure 4. Numbers from 1 to 60 in the Babylonian System

Source: https://euclides59.files.wordpress.com/2012/05/1280px-babylonian numerals- svg.png?w=772&h=464

As Gonzalez Urbareja (2008) indicates, in figure 3 it can be noticed that the symbols on the horizontal diagonal of the tablet represent the Arabic numerals: 1;24, 51, 10, where the semicolon separates the integer part from the fractional part. In the following, the procedure used to convert from Babylonian numbers to numbers of the decimal numeral system will be shown:

$$1; 24, 51, 10 = 1 + \frac{24}{60} + \frac{51}{60^2} + \frac{10}{60^3} \cong 1,4142 \ldots \cong \sqrt{2}$$

The diagonal in a square divides it into two congruent right triangles, the length of the sides can be used to find the measure of the hypotenuse. Considering L as the length of the sides of the triangle and D as the length of the hypotenuse (diagonal of the square), it is now known from the Pythagorean Theorem that the following expressions are verified:

$$L^2 + L^2 = D^2$$

$$2. L^2 = D^2$$

$$\sqrt{2}.\ \sqrt{L^2} = \sqrt{D^2}$$

$$\sqrt{2}.\ L = D$$

The diagonal is obtained by multiplying the square root of two by the length of the side. It is remarkable how the Babylonians used this property because if you look at the tablet in figure 2 at the bottom the numbers that appear are: 42; 25, 35, which represent in our current system:

$$42;\ 25, 35\ = 42 + \frac{25}{60} + \frac{35}{60^2} \cong 42,426 \ldots$$

You can see that if you multiply 30 by 1.4142. you get 42.426... which is the length of the diagonal.

Another tablet found was the PLIMPTON 322 tablet, used as a slide rule.

trigonometric, but without angles, probably for the construction of temples and buildings.

Figure 5. Cuneiform Tablet

Source: *Columbia University Libraries Online Exhibits*, accessed 14 June 2020. Available at https://exhibitions.library.columbia.edu/exhibits/show/jewels/item/10987

The PLIMPTON Tablet shown in figure 5 seems to scientists to be at first a mere commercial record of the Babylonians. On the website of the University of Columbia you can see the tablet which has dimensions of 13 x 9 cm, and a thickness of 2 cm. This tablet consists of fifteen rows and four columns, only numbers, some of which cannot be seen, however, once the *formation law* of the tablet was obtained, it was possible to calculate them. Figure 6 shows the numbers written with Arabic numerals. While in figure 7, as it was done before with the YBC 7289 Tablet, they are presented transformed to the current decimal system.

C1								C2			C3			C4
1	59	0	15						1	59		2	49	1
1	56	56	58	14	50	6	15		56	7	1	20	25	2
1	55	7	41	15	33	45		1	16	41	1	50	49	3
1	53	10	29	32	52	16		3	31	49	5	9	1	4
1	48	54	1	40					1	5		1	37	5
1	47	6	41	40					5	19		8	1	6
1	43	11	58	28	26	40			38	11		59	1	7
1	41	33	59	3	45				13	19		20	49	8
1	38	33	36	36					8	1		12	49	9
1	35	10	2	28	27	24	26	1	22	41	2	16	1	10
1	33	45								45		1	15	11
1	29	21	54	2	15				27	59		48	49	12
1	27	0	3	45					2	41		4	49	13
1	25	48	51	35	6	40			29	31		53	49	14
1	23	13	46	40						56		1	48	15

Figure 6. Table YBC 7289 in Arabic numerals

Source: https://i0.wp.com/cienciaxxi.es/blog/wp-content/uploads/2009/02/plim2.jpg

Interpretación (decimal)				b=c/raiz(C1)
$(c/b)^2$	a	c		
C1	C2	C3	PRUEBA	
1,98340278	119	169	1,98340278	120
1,94915855	3367	4825	1,94915855	3456
1,91880213	4601	6649	1,91880213	4800
1,88624791	12709	18541	1,88624791	13500
1,81500772	65	97	1,81500772	72
1,7851929	319	481	1,7851929	360
1,71998368	2291	3541	1,71998368	2700
1,69277344	799	1249	1,69270942	960
1,64266944	481	769	1,64266944	600
1,58612257	4961	8161	1,58612257	6480
1,5625	45	75	1,5625	60
1,48941684	1679	2929	1,48941684	2400
1,45001736	161	289	1,45001736	240
1,43023882	1771	3229	1,43023882	2700
1,38716049	56	106	1,38716049	90

Figure 7. Table YBC 7289 in current decimal system
Source: https://i2.wp.com/cienciaxxi.es/blog/wp-content/uploads/2009/02/plim3.jpg

To explain the preceding tables, consider a right-angled triangle such as the one in the figure 8:

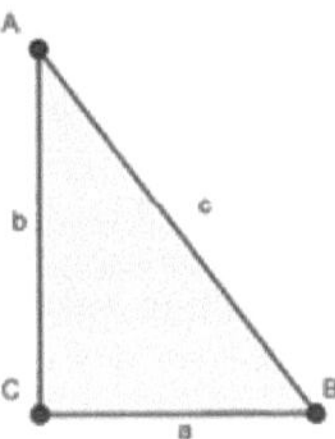

Figure 8. Rectangular Triangle

Source: Own elaboration with GeoGebra software.

21

Where **a** is the minor leg, **b** is the major leg and **c** is the hypotenuse.

In figure 6, according to Gonzalez Urbareja (2008), the fourth column shows numbers from 1 to 15 and indicates the row number, the second and third columns show the values of **a** (minor leg) and **c** (hypotenuse) respectively. It could be thought that this table expresses the values of the squares of the reason **c** over **b**, that is, the squared cosecant of the angle B (formed by the major leg and the hypotenuse) of the right triangle ABC in figure 8.

(But how were they found? It is suspected that the scribes began to make the table by taking two regular sexagesimal integers, i.e. numbers that must be divisible by 2, 3 and 5 (given that the system is sexagesimal and 60 is divisible by these only three prime integers). Gonzalez Urbareja (2008) indicates that these two numbers were called u and v, as explained above, the tablet shows the values of **b** and **c**. The specialists established that these values were found considering u > v, as follows:

$$a = 2uv, b = u^2 - v^2 \ y \ c = u^2 + v^2 \quad (1)$$

By the Pythagorean theorem it is known that:

$$c^2 = a^2 + b^2 \quad (2)$$

Substituting the equalities (1) into (2), it is found that the Pythagorean theorem is verified.

$$c^2 = a^2 + b^2$$

$$(u^2 + v^2)^2 = (2uv)^2 + (u^2 - v^2)^2$$

$$u^4 + v^4 + 2u^2v^2 = 4u^2v^2 + u^4 + v^4 - 2u^2v^2$$

$$u^4 + v^4 + 2u^2v^2 = u^4 + v^4 + 2u^2v^2$$

the Pythagorean Theorem is verified

The PLIMPTON 322 tablet presents the solutions of the Pythagorean equation $c^2 = a^2 + b^2$, whose solution is not unique, since there are infinites, which are called Pythagorean triads. The Babylonians did not know the concept of infinite, so they used the formula for $1 \leq v < u \leq 60$, thus obtaining 38 Babylonian solutions shown in Figure 9. Two scientists from the University of New South Wales in Sydney, Daniel Mansfield and Norman Wildberger, claim two centuries after their discovery, that the table rethinks some things that have been known about the specialty: "In some respects it is superior even to our modern trigonometry", says Mansfield (quoted in Ramos, 2017).

These scientists consider the tablet to be a trigonometric slide rule, indeed, the oldest and most accurate one ever found. Figure 9 shows the reconstruction of the Plimpton 322 tablet with its thirty-eight rows and eight columns.

Parameters p	q	$2pq$ Height	$\left(\dfrac{p^2-q^2}{2pq}\right)^2$ Square on Width A	$\left(\dfrac{p^2+q^2}{2pq}\right)^2 = A + 1$ Square on Diagonal B	p^2-q^2 Width C	p^2+q^2 Diagonal D	No
12	5	2, 0	59, 0,15	1,59, 0,15	1,59	2,49	1
1, 4	27	57,36	56,56,58,14,50, 6,15	1,56,56,58,14,50, 6,15	56, 7	1,20,25	2
1,15	32	1,20, 0	55, 7,41,15,33,45	1,55, 7,41,15,33,45	1,16,41	1,50,49	3
2, 5	54	3,45, 0	53,10,29,32,52,16	1,53,10,29,32,52,16	3,31,49	5, 9, 1	4
9	4	1,12	48,54, 1,40	1,48,54, 1,40	1, 5	1,37	5
20	9	6, 0	47, 6,41,40	1,47, 6,41,40	5,19	8, 1	6
54	25	45, 0	43,11,56,28,26,40	1,43,11,56,28,26,40	38,11	59, 1	7
32	15	16, 0	41,33,59, 3,45	1,41,33,59, 3,45	13,19	20,49	8
25	12	10, 0	38,33,36,36	1,38,33,36,36	8, 1	12,49	9
1,21	40	1,48, 0	35,10, 2,28,27,24,26,40	1,35,10, 2,28,27,24,26,40	1,22,41	2,16, 1	10
1, 0	30	1, 0, 0	33,45	1,33,45	45, 0	1,15, 0	11
48	25	40, 0	29,21,54, 2,15	1,29,21,54, 2,15	27,59	48,49	12
15	8	4, 0	27, 3,45	1,27, 3,45	2,41	4,49	13
50	27	45, 0	25,48,51,35, 6,40	1,25,48,51,35, 6,40	29,31	53,49	14
9	5	1,30	23,13,46,40	1,23,13,46,40	56	1,46	15
16	9	4,48	22, 8,12,36,15 .	1,22, 8,12,36,15 .	2,55	5,37	16
27	16	14,24	17,58,17,38,24,10	1,17,58,17,38,24,10	7,53	16,25	17
5	3	30	17, 4	1,17, 4	16	34	18
1,21	50	2,15, 0	15, 4,53,43,54, 4,26,40	1,15, 4,53,43,54, 4,26,40	1, 7,41	2,13, 1	19
8	5	1,20	14,15,33,45	1,14,15,33,45	39	1,29	20
25	16	13,20	12,45,54,20,15	1,12,45,54,20,15	6, 9	14,41	21
3	2	12	10,25	1,10,25	5	13	22
40	27	36, 0	9,45,22,16, 0,40	1, 9,45,22,16, 0,40	14,31	38,49	23
36	25	30, 0	8,16,16, 4	1, 8,16,16, 4	11,11	32, 1	24
1, 4	45	1,36, 0	7,44,29,13, 8,26,15	1, 7,44,29,13, 8,26,15	34,31	1,42, 1	25
45	32	48, 0	7,14,53,46,33,45	1, 7,14,53,46,33,45	16,41	50,49	26
25	18	15, 0	6,42,40,16	1, 6,42,40,16	5, 1	15,49	27
27	20	18, 0	5,34, 4,37,26,40	1, 5,34, 4,37,26,40	5,29	18,49	28
4	3	24	5, 6,15,22, 3,45	1, 5, 6,15,22, 3,45	7	25	29
32	25	26,40	3,42,55	1, 3,42,55	6,39	27,29	30
5	4	40	3, 2,15	1, 3, 2,15	9	41	31
6	5	1, 0	2, 1	1, 2, 1	11	1, 1	32
32	27	28,48	1,44,55,12,40,25	1, 1,44,55,12,40,25	4,55	29,13	33
9	8	2,24	1,12,15	1, 1,12,15	17	2,25	34
10	9	3, 0	0,40, 6,40	1, 0,40, 6,40	19	3, 1	35
27	25	22,30	0,21,21,53,46,40	1, 0,21,21,53,46,40	1,44	22,34	36
16	15	8, 0	0,15, 0,56,15	1, 0,15, 0,56,15	31	8, 1	37
25	24	20, 0	0, 6, 0, 9	1, 0, 6, 0, 9	49	20, 1	38

Proposed reconstruction of Plimpton 322. *Derek J. de Solla Price*

Figure 9. Reconstruction of Plimpton Tablet 322 with its 38 Rows and 8 Columns

Source: *http://francis.naukas.com/files/2017/09/Dibujo20170906-proposed-reconstruction-plimpton-322-babylonian-clay-tablet.png*

Now, in the above examples it can be seen that the Babylonians left evidence that they used the "Pythagoras Theorem", but there is no evidence about the elaboration of any statement of it, much less its proof. This leads one to think that they performed a *pragmatic proof of a generic model*, but without a statement of the theorem. Would the same have happened with the Egyptians?

3.2 The Pythagorean Theorem in Ancient Egypt

It is not known exactly if the ancient Egyptians knew the "Pythagoras Theorem" for any three sides of a right triangle, what is known is that they knew what they called the *Egyptian Sacred Triangle*. It was a rectangular triangle whose sides measured 3, 4 and 5. Each of them represented a divinity; the side measuring 3 units represented Osiris, the side measuring 4 represented Isis and the side measuring 5 represented Horus. This triangle was used to form right angles which were then applied, for example, in the construction of pyramids and monuments, since they could double, triple and much more to these measures. A rope was used with 12 equidistant knots two by two, each of the distances between each knot representing 1 unit. See Figure 10. *Egyptian sacred triangle* (Gonzalez Urbareja, 2008).

$$3^2 + 4^2 = 5^2$$

$$6^2 + 8^2 = 10^2$$

Figure 10. Egyptian sacred triangle

Source: http://wordpress.colegio-arcangel.com/matematicas/files/2012/10/13-teorema.jpg

Most of the mathematical knowledge in Egypt comes from two substantial papyri, each named after its ancient owner. One is the Rhind papyrus, purchased in 1858 by the Scottish Egyptologist Alexander Henry Rhind. This papyrus uses the hieratic script (a cursive form of hieroglyphics better adapted to the use of pen and ink) circa 1650 BC by a scribe named Ahmes, who claimed it was the likeness of an earlier work dating to the Twelfth Egyptian Dynasty, 1849-1801 BC (Burton, 2007, p.34). The Ahmes or Rhind Papyrus can be seen in Figure 11.

Figure 11. Rhind Papyrus

Source: British Museum Images, accessed 24 June 2020. Available at: https://www.bmimages.com/preview.asp?image=00569564001&itemw=4&itemf=0004&itemstep=1&itemx=4

On the website of the British Museum you can find the papyrus and the following description: Several documents have survived that shed some light on the mathematical approach of the ancient Egyptians. The best known and longest is the Rhind Mathematical Papyrus, acquired by the Scottish lawyer A.H. Rhind in Thebes around 1858. Budge's original introduction to the papyrus facsimile indicates that these fragments were found in a chamber of a ruined building near the Ramesseum. The two sections in the British Museum were joined by a now missing section about 18 cm long; the original may have been cut in half by modern thieves to increase its sale value. The fragments that partly fill this gap were identified in 1922, in the collection of the New York Historical Society, which acquired them from Edwin Smith. Smith also acquired a chirurgical papyrus of approximately the same date as the Rhind papyrus, suggesting that these two documents may have come from a cache of early New Kingdom manuscripts (British Museum, 24 June 2020).

The papyrus is probably a mathematical textbook, used by scribes, who were the main literate section of the population, to learn how to solve particular mathematical problems by writing down appropriate examples. The text includes eighty-four arithmetical problems: division tables, multiplication and fraction handling; geometry problems, including volumes and areas; and miscellaneous problems. Of these problems, we can highlight problem 51, where the area of an isosceles triangle is sought. This is done by taking half the base of 4 jets and multiplying it by the height of 10 jets. The "jet" or "rod", was a measure of 100 cubits, and the Egyptian royal cubit equals 52.3 cm (it was divided into 7 palms). It is believed that Ahmes justified this procedure because he considered the isosceles triangle as one made up of two rectangular triangles (Boyer, 1986; Alsina, 2010).

The other is the Golenischev papyrus, after the Egyptologist Vladimir Golenischev who acquired the document in Thebes in about 1893, and later took it to the Pushkin Museum of Fine Arts in Moscow, where it remains today.

The fact that we know that they work with right-angled triangles and that they used the Egyptian sacred triangle for constructions does not imply that they have formalised the theorem, nor is there any evidence for it.

This leads one to think that the Egyptians performed a *pragmatic proof of the generic model*, but without a statement of the theorem.

3.3 The Pythagorean Theorem in Ancient India

According to Gonzalez Urbareja (2008), in India, due to the need to build temples, arithmetical and geometrical knowledge was developed between the eighth and second century BC. Amongst the Sulvasutras (people dedicated to string tensioning) the Pythagorean triads were used, such as (3, 4, 5); (5,12,13); (8,15,17); (7,24,25). They called the rectangular triangle with sides 5, 12 and 13 the "Indian triangle". They also classified the triads according to whether the difference between the measure of the hypotenuse and the measure of the major leg is 1, 2 or 3. As shown in Figure 12:

$c - b = 1$			$c - b = 2$			$c - b = 3$		
a	b	c	a	b	c	a	b	c
3	4	5	8	15	17	15	36	39
5	12	13	12	35	37			
7	24	25						

Ternas pitagóricas de los hindúes

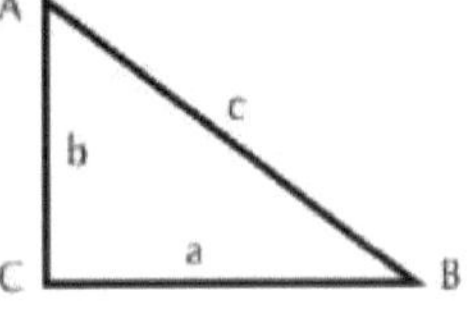

Figure 12. Hindu Pythagorean Ternas

Source: http://www.hezkuntza.ejgv.euskadi.eus/r43-573/en/content/information/information/dia6 sigma/en sigma/adjuntos/sigma 32/8 pitagoras.pdf

"The fact that all the Pythagorean terns that appear in the Sulvasutras can be easily derived from the old Babylonian rule for their construction, allows us to assure the Mesopotamian influence on the Hindu knowledge on the subject" (Gonzalez Urbareja, 2008, p.109.) This is demonstrated in Figure 13, which shows the traces of the trapezoidal altars of the Sulvasutra of Apastamba (5th century B.C.).

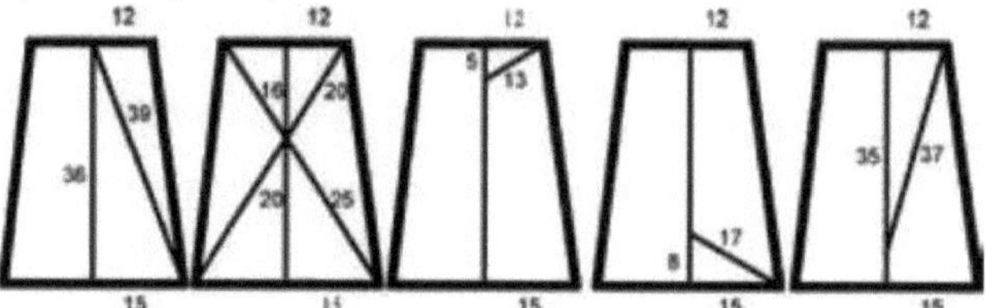

Figure 13. Traces of the trapezoidal altars of the Sulvasutra of Apastamba (5th century BC) with indication of the Pythagorean terns used in the ritual construction.

Source: Gonzalez Urbareja, 2008, p.109

This leads one to think that, in Ancient India, *pragmatic proofs of the generic model* were made, but without a statement of the Pythagorean Theorem.

3.4 **The Pythagorean Theorem in Ancient China**

In ancient China the Pythagorean Theorem is known as Kon Ku or Gougu. There are two classical Chinese treatises of mathematical content where geometric aspects linked to the Pythagorean Theorem are related, they are the Chou Pei Suan Ching (300 B.C.) and the Chui Chang Suang Shu (250 B.C.).

In the Chou Pei there is first a figure called the "Hypotenuse Diagram", which can be seen in Figure 14.

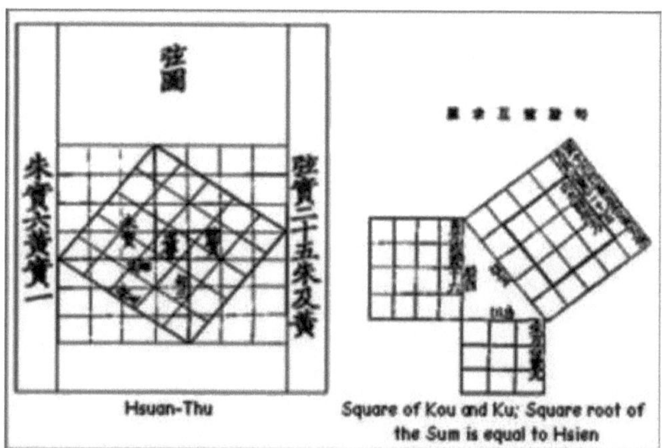

Figure 14. Pythagoras theorem in the Chou Pei

Source: Il teorema Kou Ku secondo l'illustrazione orginale del Chou Pei riprodotta in Needham 1959, p. 22.

Needham (1959) sets out an explanation for such a test, which is clarified by considering Figure 15.

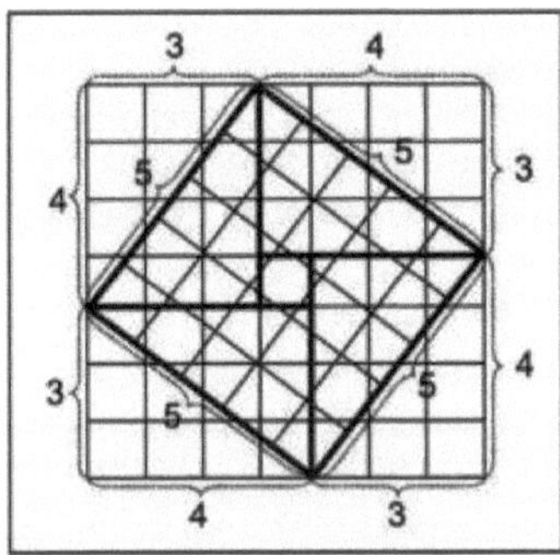

Figure 15. "Hypotenuse diagram".

Consider a rectangle with a base equal to 4 units and a height of 3 units, it is enlarged in such a way as to obtain a square with a side equal to 7 units. The diagonal between the two "corners" will be 5 units long. Then draw a square whose side is equal to the value of this diagonal. It is circumscribed by half rectangles like that, which has been left out, to form a (square) plate. Thus, the (four) outer half-rectangles of width 3, length 4 and diagonal 5 form two rectangles (of area 24); then (when this is subtracted from the square plate of area 49) the remainder is of area 25. This (process) is called "stacking the rectangles" (Needham, 1959, pp. 22-23).

"On the other hand, the Chui-Suang contains 246 problems of which 24 refer to right triangles. All the solutions to the problems are based in one way or another on the Pythagorean Theorem" (Acevedo, 2011, p.12.). The most famous is the broken bamboo problem with which the mathematicians Chao Ching Ching and Liu Hiu proved the result of the Pythagorean Theorem for any right triangle, as shown in Figure 16. The problem stated the following:

"There is a bamboo ten feet high, which has broken in such a way that its upper end rests on the ground at a distance of three feet from the base. You are asked to calculate at what height the break has occurred".

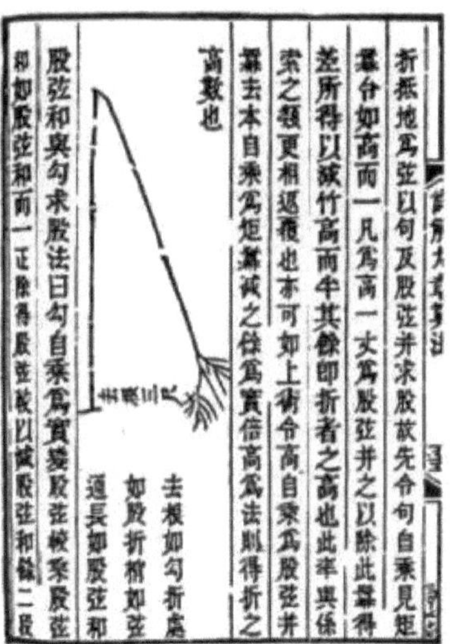

 Chinese Broken Bamboo Problem
Source: http://elcuadradodelahipotenusa.blogspot.com/p/el-teorema-de-pitagoras-en-la-india-y.html

This problem combines the Pythagorean Theorem with the solution of quadratic equations, since the solution requires solving the equation: $x^2 + 9 = (10-\%)^2$ (Gonzalez Urbareja, 2008, p.109).

Another example is Figure 17, where we can observe the *Intellectual Test of Visual Support*, which according to Alsina (2010) is found in the Chou Pei Suan Ching.

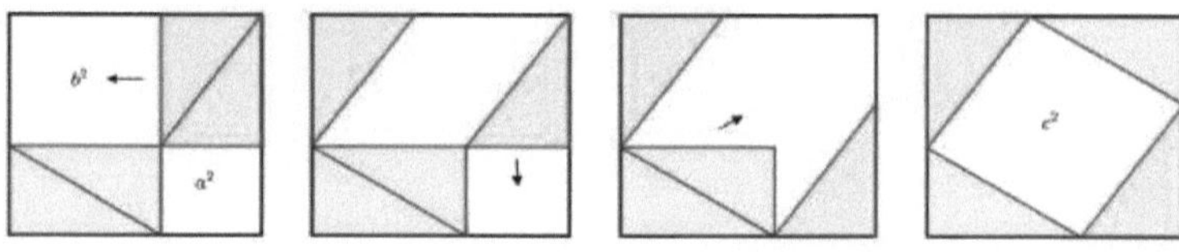

Figure 17. A Pragmatic Test of the Generic Chou Pei Suan Ching Model
Source: Own elaboration with GeoGebra software according to Alsina (2010).

Below, in Figure 18, are two pages from the *Zhoubi Suanjing (Treatise on the Nine Chapters of the Mathematical Art),* as Swetz and Katz indicate, these images are from a Ming dynasty copy printed in 1603. These diagrams were added to the original text at some point in an attempt to illustrate a dissection proof of the 'Pythagoras Theorem' (Swetz and Katz, 2011, para.1).

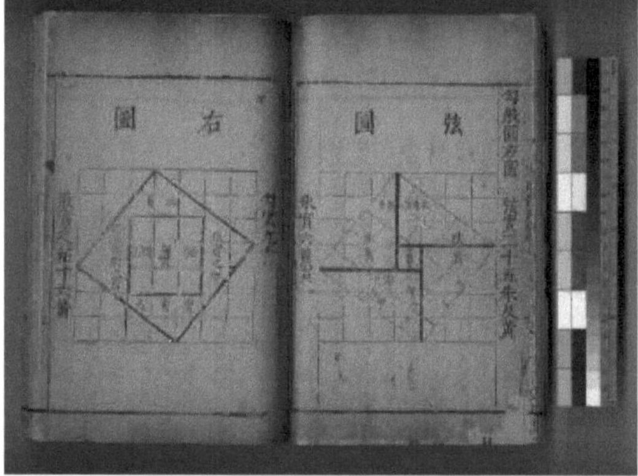

Figure 18. Chinese Zhoubi Suanjing Book

Source: https://www.maa.org/sites/default/files/images/cms_upload/0800807640911.jpg

In these pages, the diagram on the right is usually called the "hypotenuse diagram" and illustrates the proof of *Gougu's* (or Pythagorean) theorem in the 3-4-5 case. The diagram on the left shows how a square of side 3 fits into a square of side 5 (Swetz and Katz, 2011, parr.2).

The diagram in Figure 19 illustrates a square whose side measures 4, fitting into a square whose side measures 5. In Figure 20 made in GeoGebra, you can see the decomposition of the squares in Figure 19 into rectangular triangles.

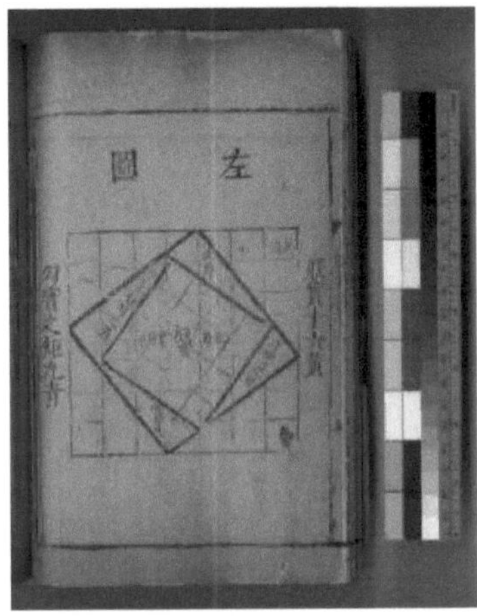

Figura 19.   Chinese Zhoubi Suanjing Book
Source: https://www.maa.org/sites/default/files/images/cms_upload/0800807743466.jpg

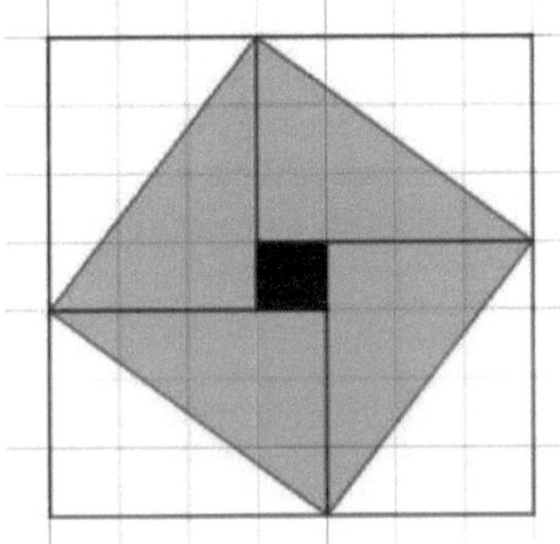

Figura 20.   Side 4 Square Fitting into Side 5 Square according to Chinese Book Zhoubi Suanjing
Source: Own elaboration with GeoGebra software.

The tests of the ancient Chinese civilisation can be classified as *Intellectual tests with visual support,* according to the Taxonomy of Tests, developed in section 2.2.

3.4 The Pythagorean Theorem in Ancient Greece

3.4.1 Pythagoras of Samos

Pythagoras' father was Mnesarchus, while his mother was Pythais and a native of Samos. Mnesarchus was a merchant who came from Tyre, and the story goes that he was the one who brought corn to Samos in a time of famine and was rewarded with the city of Samos as a sign of gratitude. As a child, Pythagoras spent his early years on Samos, but travelled widely with his father. There are chronicles of Mnesarchus returning to Tyre with Pythagoras, and he was taught by the Chaldeans and by the wise men of Syria. It seems that he also visited the present region of Italy with his father (Aznar, 2007, para. 4).

Figura 21.   Erma of "Pythagoras". Marble Sculpture
Source http://www.museicapitolini.org/es/mostre ed eventi/eventi

It is said that Pythagoras visited Thales of Miletus when he was between 18 and 20 years old. Thales was already an old man and it was Thales who suggested that he travel to Egypt to learn more about mathematics and astronomy (Aznar, 2007). It is estimated that Pythagoras went to Egypt in about 535 B.C. He may have learned geometrical notions from Thales and his disciples, but also from the teachings of the high priests in Egypt. When the king of Persia invades Egypt, Pythagoras is taken prisoner and sent to Babylon, where he may have come into contact with the advances of this civilization in mathematical notions. Some time later he was released. It was many more years before he founded his school in Crotona, in southern Italy.

There are several stories about his death, but what is known is that he died around 500 BC (Alsina, 2010; Aznar, 2007).

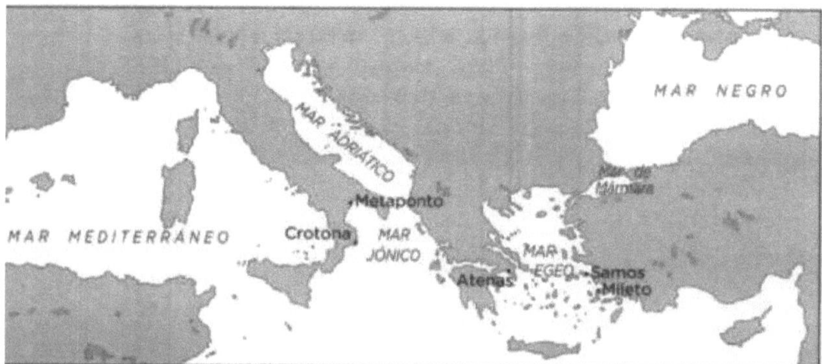

Figure 22. Map showing Athens, Miletus and the main cities related to the biography of Pythagoras.

Source: taken from Alsina (2010).

In Pythagoras' time, the Theorem had an immediate and very clear practical use which follows from the first and most essential of its properties: it served to determine perpendicularities (Jaen Sanchez, 2012, p.37).

A very persistent tradition with documentary basis in Vitruvius, Plutarch, Diogenes Laertius, Athenaeus and Proclus, attributes the Pythagorean Theorem to Pythagoras himself. Vitruvius writes in The Ten Books of Architecture (Book IX, Ch.2), quoted by Gonzalez Urbareja (2008, p.24):

"Pythagoras found and proved theoretically the form of the square, and by his reasoning and method obtained a perfect square: a coda which the artificers, after much labour, can scarcely achieve. For if you take three rulers, a long one of three feet, another of four, and the third of five, adapting them in such a way that they touch each other at their extremities in the shape of a triangle, you will have a perfect square. [...] When Pythagoras found this, not doubting that the muses had enlightened him in his invention, they say that he made sacrifices to them in thanksgiving".

It is not known exactly what the proof given by Pythagoras or the Pythagoreans was, but one possible proof is thought to be that found in Euclid's Elements:

Let it be the right triangle in Figure 23, draw on its sides, squares as shown in Figure 24.

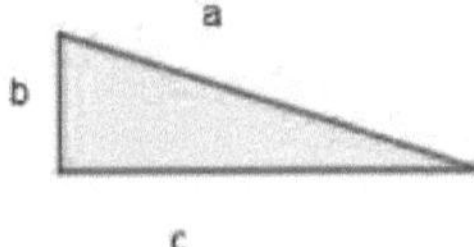

Figure 23. Rectangular Triangle

Source: Own elaboration with GeoGebra software.

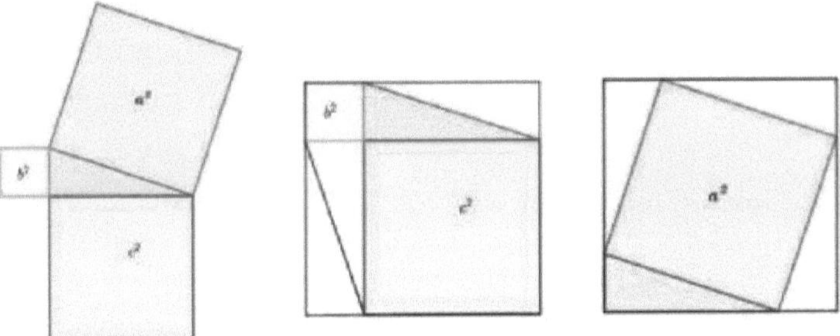

Figure 24. Test performed by Pythagoras of Samos

Source: Own elaboration with GeoGebra software based on the text of Gonzalez Urbareja (2008).

Then, on the square with side (b + c), the figures are rearranged, first the squares with sides b and c are placed together with the pink triangle, as can be seen in the centre of Figure 24. While in the figure on the right the square with side **a** and the pink triangle are ordered, this allows to verify that: $a^2 = b^2 + c^2$, since three white triangles and one pink one are obtained inside each square with side (b + c), therefore it is verified that the areas of the orange and blue triangles are equivalent to the green one.

This test can be compared to the test carried out in the Chinese Zhoubi suanjing book (Gonzalez Urbareja, 2008).

Another possible proof of this Theorem, given by Pythagoras, could be the following:

Consider the following right triangle ABC in Figure 25:

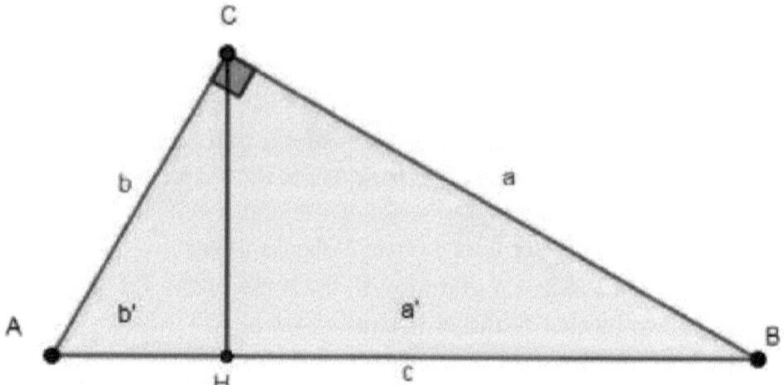

Figure 25. Proof by Pythagoras of Samos

In the said rectangular triangle ABC at C; the segment CH is the height relative to the hypotenuse, which separates the segment c into the segments a' and b', which correspond to the orthogonal projections on c, of the legs a and b, respectively. The right triangles *AHC and CHB,* which are similar to each other and to the triangle ACB, are then determined. It can be proved in the following way:

- Test of the similarity between *CBA* and *AHC:*

Two triangles are similar if they have two or more congruent angles (as in Euclid's Elements):

$$\frac{b}{b'} = \frac{c}{b}$$

$$b^2 = b'c \quad (1)$$

- Evidence of the similarity between *CBA* and *CHB:*

$$\frac{a}{a'} = \frac{c}{a}$$

$$a^2 = a'c \quad (2)$$

Adding (1) to (2) gives:

$$b^2 + a^2 = b'c + a'c = c.(b' + a')$$

But as $(a' + b') = c,$, it finally turns out

$$b^2 + a^2 = c^2$$

Therefore, these triangles are similar (Jadn, 2012).

This proof is anachronistic, since the Greeks did not use algebra to prove, but it is used in this work to explain Figure 25, considered a possible proof made by Pythagoras. Both proofs are of the *Intellectual* type *of formal and geometrical demonstration.*

3.4.2 **Pythagoras' Theorem according to Plato**

Plato was a Greek philosopher who lived between 427 B.C. and 347 B.C. He was born in Athens, Greece.

Figure 26. Bust of Plato from the 4th century BC.
Source: https://upload.wikimedia.Org/wikipedia/commons/thumb/d/da/Plato Pio-
Clemetinolnv305.ipg/800px-PlatoPio-Clemetinolnv305.jpg

In the school of Plato founded about 387 BC (Figure 27), the particular case of the Pythagorean Theorem for the isosceles right triangle was developed.

Figure 27. School of Athens, Fresco by Raphael (1509-1510)

Source: http://m.museivaticani.va/content/museivaticani-mobile/es/collezioni/musei/stanze-di-
raffaello/stanza-della-segnatura/scuola-di-atene.html

This test begins by considering a red square, as shown in Figure 28. Its sides measure two units and its area is worth 4 square units, so that by drawing a new square like the blue one on its diagonal, a square of 8 square units is obtained, which has twice the surface of the first one (Gonzalez Urbareja, 2008). It can be seen that the area of the blue square is equal to the sum of the areas of the red squares.

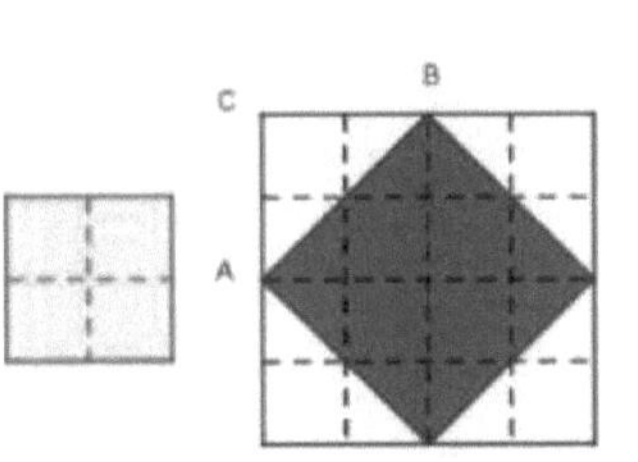
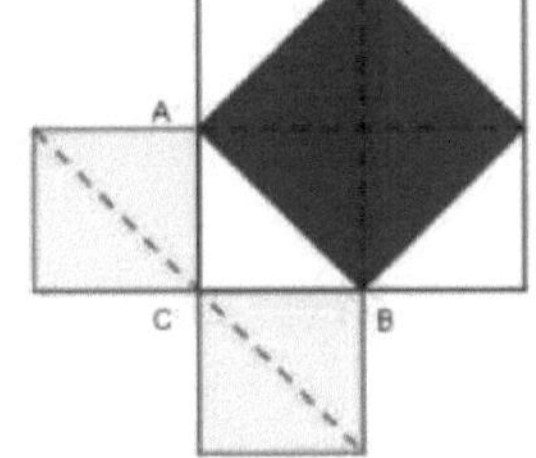

Figure 28. Rectangular Triangle Isosceles of Plato

Plato also concludes it by using the laws that serve to construct Pythagorean triads, in which the difference between the measure of the hypotenuse (c) and the measure of the vertical leg (b) or height is equal to two. The laws of construction are the following:

$$a = 2m$$

$$b = m^2 - 1 \quad y \quad c = m^2 + 1,$$ with natural m.

The idea of these laws is derived from the Babylonians, the Hindus and the Pythagoreans themselves (Kline, 1994, cited by Hernandez Cruz, 2019).

In Plato's work, "The Menon", Socrates asks a slave and from his answer appears the problem of the duplication of the square, which in time derives in the problem of the duplication of the cube (Gonzalez Urbareja, 2008; Hernandez Cruz, 2011).

According to the taxonomy presented in section 2.2, this test is classified as a *visually supported intellectual test*.

3.4.3 Pythagoras' Theorem according to Euclid

Euclid was a Greek mathematician who lived approximately between 330 BC and 275 BC, born in Alexandria, and is recognised as the father of geometry. It is said that he studied in the school of Plato and then created his own school in Alexandria. He produced numerous works including his famous work "The Elements of Geometry" (Alsina, 2010, Jadn, 2012).

Figura 29.   Statue of Euclid. Oxford University Museum of Natural History
Source: https://mateturismo.wordpress.com/tag/euclides/

Euclid is supposed to have produced this work in Alexandria. The original work is unknown, but throughout history there are numerous writings based on his translations. At first it is believed that Teon and his daughter Hypatia made the first translation of the manuscript, which was later translated into Arabic and Latin (Alsina, 2010; Boyer, 1986; Jadn, 2012).

Some of these translations are presented below. Figure 30 shows a manuscript in Greek, one of the earliest known editions of "The Elements".

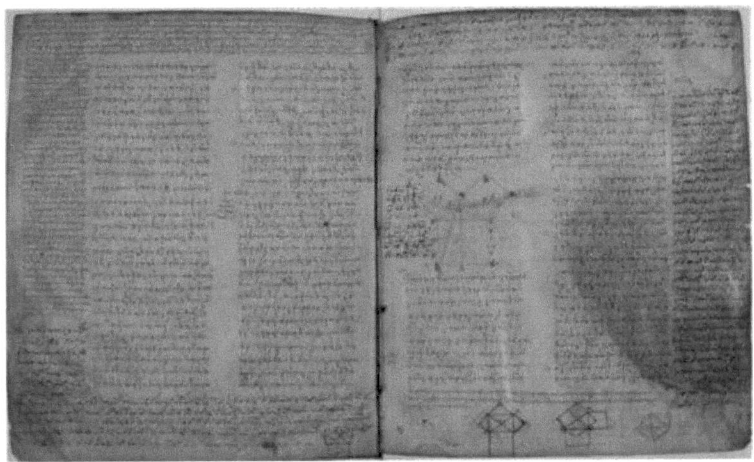

Figure 31 shows an Arabic version by the Persian scholar Nasir al-Din Muhammad ibn Muhammad al-Tusi (1201-1274).

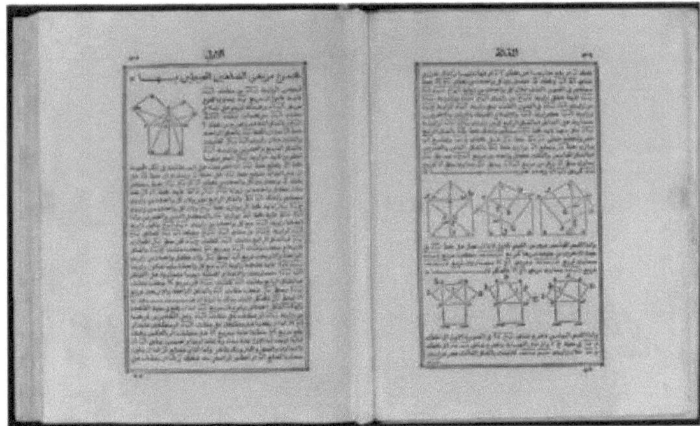

Figura 31.   The Recension of Euclid's Elements by the Persian Scholar Nasir al-Din Muhammad ibn Muhammad al-Tusi (1201-1274)

Figure 32 shows the first printed edition of the most famous work of Euclid, appeared in Venice in 1482, translated from Arabic into Latin.

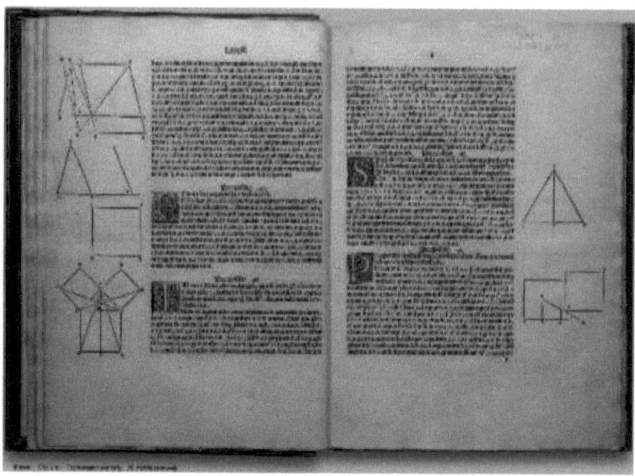

Figura 32.   Printed edition of Euclid's most famous work. Erhard Ratdolt, 1482
Source: http://www.rarebookroom.org/Control/eucgeo/index.html

Figure 33 shows the title page of the first Spanish edition of "Euclid's Elements", which dates from 1576.

Figura 33.   Cover of the First Edition in Spanish Language of Euclid's Elements (Rodrigo Qamorano, Seville, 1576)
Source:
https://ichef.bbci.co.uk/news/ws/660/amz/worldservice/lrve/assets/images/2016/05/10/160510184233_euclides-en-en-espanol.jpg

The Greek philosopher Proclus states, at the end of his brief catalogue of Eucbidean works, that "The Elements have an incontestable and perfect grammar of the scientific exposition itself in the matter of geometry" (Euclid, trans. in 1991).

The Elements is composed of 13 books on geometry, arithmetic and algebra; it contains 467 theorems on plane geometry (books I to IV), proportion theory (books V to VI), number theory (books VII to X) and geometry of space (books XI to XIII). Each book has definitions and theorems except the first book which also contains 5 postulates and 5 common notions or axioms. The axioms are chosen as self-evident truths common to all sciences. The postulates are chosen as self-evident truths specific to a particular science, in this case geometry (Sanchez, 2012, p. 78).

The first Book I of Euclid's Elements ends with the most important theorem of elementary geometry: The Pythagorean Theorem and its reprocal (Propositions I.47 and I.48).

Euclid states the Pythagorean Theorem as follows:

"In right-angled triangles the square of the side that subtends the right angle is equal to the squares of the sides that comprise the right angle" (Euclid, trans. in 1991, p.260).

Not being able to use the proportions in Pythagorean form -because of the presence of incommensurable magnitudes- that suppose the application of similarity -which does not appear in The Elements until Book VI-, Euclid sharpens his wits and obtains the magnificent result by applying very simple elements of elementary Geometry, previously studied, for his demonstration. This demonstrates the importance of contextualising knowledge according to the knowledge possessed up to that moment, and in the classroom it is fundamental to do so.

Euclid uses to demonstrate this proposition (Gonzalez Urbareja, 2008, p. 114):

- The construction of squares on segments (I.46).
- Adjacent angles that add up to two straight lines (I.14).
- The first triangle congruence theorem (I.4).
- The relationship between triangles and parallelograms having the same base and lying between the same parallels (I.36, I.41).

"Parallelograms having the same base and lying between the same parallels have the same area" (Elements, I.36).

"If a parallelogram has the same base as a triangle and they are situated between the same parallels, the area of the parallelogram is twice that of the triangle" (Elements, I.41).

It is therefore a clear example of the application of an axiomatic system for the proof of the truth of a theorem.

Demonstration of Proposition I.47 (Pythagoras' Theorem):

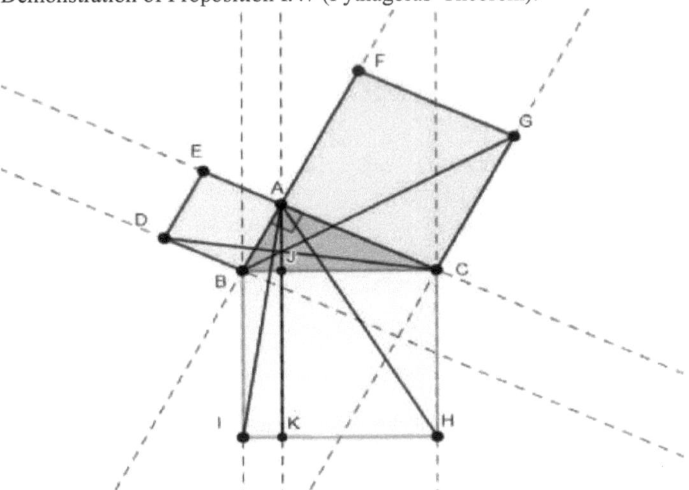

Figure 34. Demonstration of Proposition I.47

Euclid uses propositions I.36 and I.47. for the proof of the theorem:

i. Triangles BDC and ABI are equal because *AB = BD, BI = BC* and angle B of triangle

DCB is equal to angle B of triangle ABI.

ii. The area of the green square ABDE is twice the area of the triangle BDC since they share the side *BD* and are located between the same parallels.

iii. The area of the rectangle BIKJ is twice the area of the triangle ABI since they share the side *AB* and are located between the same parallels.

From dem i. ii. and iii., it follows that the area of the rectangle BIKJ is equal to the area of the square ABDE.

Analogously it is shown that the area of the rectangle CHKJ is equal to the area of the blue square GFAC.

Then, since the area of the orange square BIHC is equal to the sum of the areas of the rectangles BIKJ and CHKJ.

Thus the area of the orange square BIHC, outer square on the hypotenuse, is equal to the sum of the areas of the green square ABDE and the blue square ACGF, outer squares on the legs (Acevedo, 2011).

Proposition I.48 is the reproach of proposition I.47 and states that:

"If in a triangle the square built on one of the sides is equal to the squares built on the remaining sides of the triangle, the angle comprised by those remaining sides of the triangle is right" (Euclid, trans. in 1991, p.263).

c

*

Figura 35.   Demonstration of Proposition I.47
Source: Own elaboration with GeoGebra software based on the text Acevedo (2011).

Demonstration of Proposition I.48:

In this demonstration, Euclid draws a segment |ЛВ| = |ДВ| and perpendicular to the segment AC.

From the hypothesis: $(lAlTl)^2 + (ЖТ)^2 = (|BC|)^2$,

As the triangle ADC is rectangular, the result is: $(JAD|)^2 + (|ЛC|)^2 = (|BC|)^2$, (I.47, Pythagoras' Theorem).

But since |AB| = |AD|, it will be:

$$\backslash BC|^2 = |AB \mid |^2 + |ДC|^2 = |AD|^2 + |ДC|^2 = |BC|^2$$

Therefore \BC | = |JDC|; so that the triangles DAC and CAB are equal, since the AC side is common, the two triangles have the three equal sides.

In this case, the proof is of an *intellectual type, of formal geometric and algebraic demonstration,* since it is based on the comparison of areas and on relations between sides and segments, respectively. It should be clarified that the algebraic proof is analogical to the time and is carried out by way of explanation.

A curiosity and application is the Bride's Chair diagram, often referred to as the windmill or peacock's tail, shown in Figure 36.

Figura 36.   Diagram of Theorem I.47 of Euclid's Elements Called "Bride's Chair" Source: Boyer p.150

Euclid also presents a graphical approach to the problem (Figure 37), considering the same initial triangle and transforming each square on a leg into its parallelogram of equal area (since they have equal base and equal height), and then repositioning these parallelograms in the square on the hypotenuse. This ingenious test clarifies the part occupied by each of the squares on the legs (Jadn Sanchez, 2012).

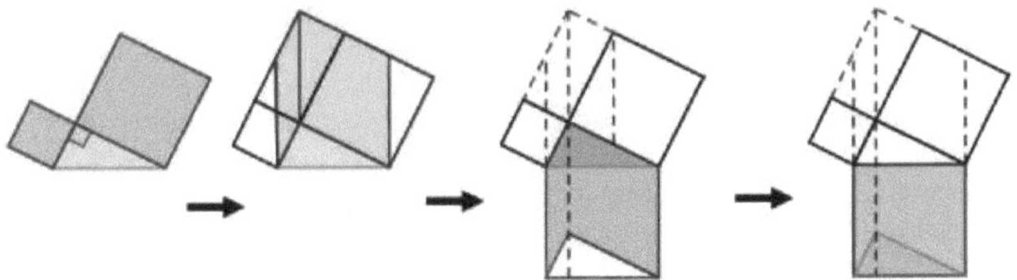

Figura 37.   Euclid's Graphical Approximation of the Pythagorean Theorem
Source: Own elaboration with GeoGebra Software based on the text Jaen. p.52

In this case the proof is of the *intellectual type of formal and geometrical demonstration.*

3.4.4 Pythagoras' Theorem according to Pappus

Pappus of Alexandria (284 - 305 AD) was a great mathematician who wrote Synagoge (Collection) around 320 AD, which compiles theorems, demonstrations, alternative lemmas of other mathematicians, such as Arqrnmides, Apollonius, Euclid to name a few. It also provides new discoveries and generalisations.

Pappus' proof is similar to that performed by Euclid, it is based on the comparison of areas of figures of the same base, located between parallels. See Figure 38.

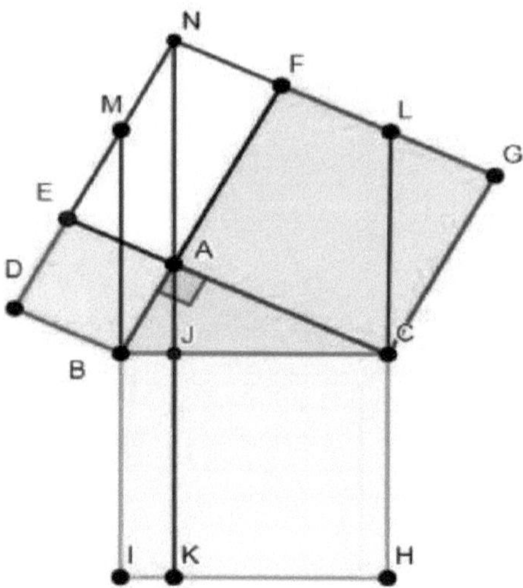

Figure 38. Pappus' Proof of the Pythagorean Theorem

Source: Own elaboration with GeoGebra software based on the text Gonzalez Urbareja (2008).

As the segments JK and AN are equal, the rectangle JKIB and the parallelogram NABM are equal to each other (Euclid I.36).

In turn this parallelogram is equal to the square ABDE.

Therefore the rectangle JKIB is equal to the square ABDE.

Analogously, the rectangle JKHC is equal to the square ACGF.

Thus the square BIHC on the hypotenuse BC is equal to the sum of the squares ABDE, ACGF, on the legs AB, AC (Gonzalez Urbareja, 2008, p. 117).

In this case the proof is of the *intellectual* type *of formal and geometrical demonstration.*

3.5 The Pythagorean Theorem in the Middle Ages

The Pythagorean Theorem is one of the theorems that has a large number of different demonstrations by the most diverse methods. One of the reasons for this is that in the Middle Ages, which began in 476 A.D., a new demonstration of the theorem was required in order to attain the rank of *Magister matheseos,* or "master of mathematics". In a certain sense, the tradition ended up turning this feat into a demonstration of the magisterium of universal knowledge (Jaen, 2012, p.54).

According to J.S.Georges (1927) the proof of this theorem represented for the first European universities an adequate objective, in mathematical aspects, to legitimately claim the title of mathematician.

"The bridge of the asses" *(Pons asinorum)* is the name given in the Middle Ages to the Pythagorean Theorem, and refers to the knowledge that separates the educated from the uneducated" (Bosch, 2007, p.23). This can be seen in the excerpt from Jules Verne's novel, From the Earth to the Moon:

-    Hooray for Edgard Poe! -exclaimed the assembly, electrified by the words of their president.

-    I have now concluded," continued Barbicane, "with these attempts, which I will call purely literary, and wholly insufficient to establish serious relations with the star of the nights. I must add, however, that some practical spirits have tried to enter into serious communication with it. Thus, some years ago a German geometrician proposed to send a commission of wise men to the steppes of Siberia. There, on vast plains, they were to establish immense geometrical figures, designed by means of luminous reflectors, among others the square of the hy-potenuse, vulgarly called "the bridge of the asses" by the French. Every intelligent being," said the geometrician, "must understand the scientific destiny of this figure. The Selenites, if they exist, will respond with a similar figure, and once communication has been established, it will be easy to create an alphabet that will allow us to converse with the inhabitants of the Moon". Thus spoke the German geometrician, but his project was not put into practice, and so far no direct link has existed between the Earth and its satellite. It is reserved for the practical genius of the Americans to enter into relation with the sidereal world. The means of achieving this is simple, easy, sure, infallible, and will constitute the object of my proposal (Jules Verne, trans. in 2004, pp.42, 43).

That is why there are many proofs and demonstrations by mathematicians and people related to the world of philosophy, art, among other professions.

Here are some of these tests.

3.5.1 Pythagoras' Theorem according to Thabit Ibn Qurra

According to Boyer (1986), Thabit Ibn Qurra was an Arab astronomer and mathematician who lived between 826 and 901 A.D. He translated and commented on treatises by Ptolemy, Archimedes, Euclid and Apollonius among others. Among his contributions we find that he worked with equations of the second degree, and developed a formula for the formation of the friendly numbers.

Thabit Ibn Qurra's proof is the reply to a letter from a friend who, knowing the particular case for an isosceles right triangle of the "Socratic proof" of the theorem - the passage of Plato in *The Menon* on the duplication of the square - asked him to communicate the proof of the general case (Gonzalez Urbareja, 2008, p. 118).

Thabit Ibn Qurra presents two proofs for the triangle ABC rectangle in B:

See Figure 39 for Thabit Ibn Qurra's Exhibit 1.

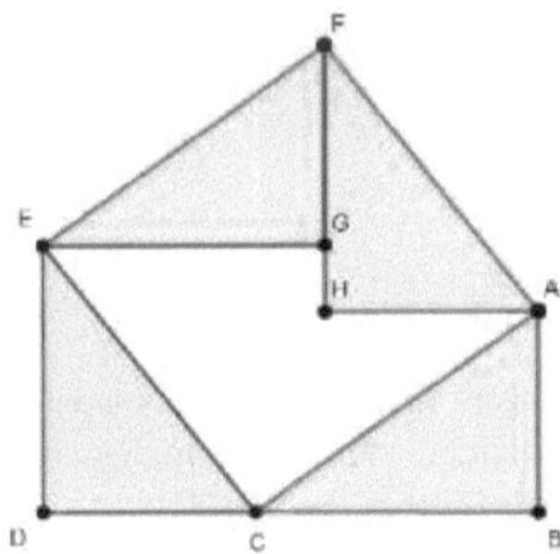

Figure 39. Thabit Ibn Qurra's Exhibit 1.

Source: Own elaboration with GeoGebra software based on the text Acevedo (2011).

Proof 1: *If the triangles ABC and CDE equal to the given one are subtracted from the figure ABCDEF, the result is the square ACEF built on the hypotenuse AC, while subtracting the triangles AHF and FGE, the result is the figure formed by the squares ABIH, GIDE, built on the legs AB, BC (Acevedo, 2011, p.27).*

Vĕase see Figure 40, for Thabit Ibn Qurra's Exhibit 2.

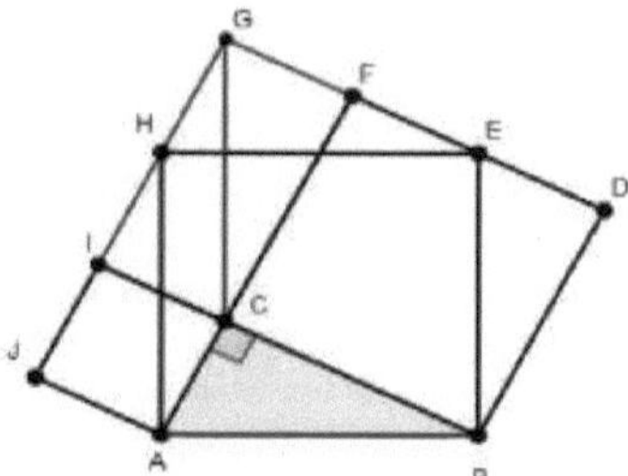

Figure 40. Thabit Ibn Qurra's Exhibit 2.

Proof 2: If we subtract from the figure ABDGJ the triangles AJH, BDE, HEG, equal to the die, we get the square ABEH on the hypotenuse AB, while subtracting the triangles ABC, CIG, CFG, we get the figure formed by the squares ACIJ, BCFD, on the legs AC, BC (Acevedo, 2011, p.27).

In this case both proofs are *Intellectual proofs of formal and geometric demonstration,* according to the Taxonomy of Proofs.

3.5.2 Pythagoras' Theorem according to Bhaskara

Bhaskara (1114-1185) was an Indian-born mathematician and astronomer, known for creating the formula for solving quadratic equations. The following is his proof of the Pythagorean Theorem, considering Figure 41.

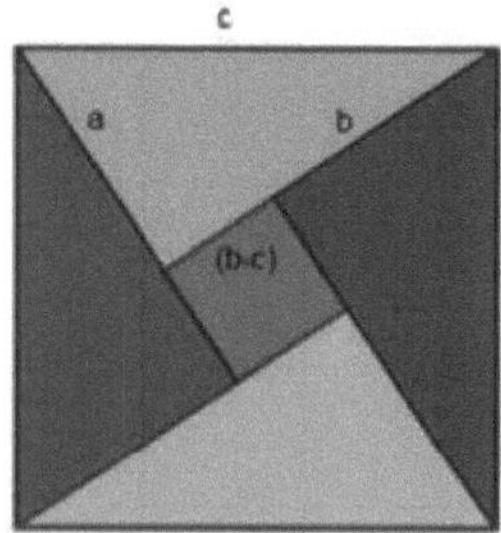

Figure 41. Bhaskara test

The square on the hypotenuse is divided into four triangles congruent to the given one and a square of side equal to the difference of the legs, that is, $b - a$. In Figure 41 on the right it can be seen that the pieces were rearranged, and the figure that turns out to be the juxtaposition of the squares on the legs was formed (Gonzalez Urbareja, 2008; Hernandez Cruz, 2019).

The geometric proof is then translated into algebraic terms by expressing the equality of the figures drawn:

$$c^2 = 4 \cdot \left[\left(\frac{1}{2} \right) ab \right] + (b-a)^2$$

$$c^2 = 2ab + b^2 - 2ab + a^2$$

$$c^2 = b^2 + a^2$$

This case is an *intellectual proof of formal and geometrical demonstration* according to the Taxonomy of Proofs.

3.6 The Pythagorean Theorem in the Modern Age

The Modern Age began in 1492, for some historians, with the conquest of America. This event brought about important changes in mathematics, related to navigation, astronomy, gravitation theory, among others. During this period geometry remained stagnant, and translations of Greek classics such as Euclid and Apollonius were made. At the end of this period, arithmetic was separated from algebra, although the latter was mainly devoted to the resolution of equations.

The proofs of the Pythagorean Theorem, carried out during this stage, by Leonardo Da Vinci and by Gopel, will be presented below.

3.6.1 Pythagoras' Theorem according to Leonardo Da Vinci

Leonardo Da Vinci lived between 1452 and 1519, was born in Italy and was an anatomist, architect, artist, writer, sculptor, philosopher, inventor, musician, among many other professions.

Figure 42. Possible Self-Portrait of Leonardo Da Vinci
Retrieved from https://historia.nationalgeographic.com. en/a/leonardo-da-vinci-ultimos-anos-genio-renaissance14142/3

Figure 43 shows one of Leonardo Da Vinci's original works, which is in the Louvre Museum in Paris, France. In addition to other contents, this work shows Da Vinci's demonstration of the Pythagorean Theorem.

Figure 43. Leonardo da Vinci's demonstration on display at the Louvre Museum. Photo: Raphael Yaghobzadeh/AP

Leonardo Da Vinci proposed a *visually supported intellectual proof* of the Pythagorean Theorem, relating the area of the figures considered, as can be seen in Figure 44:

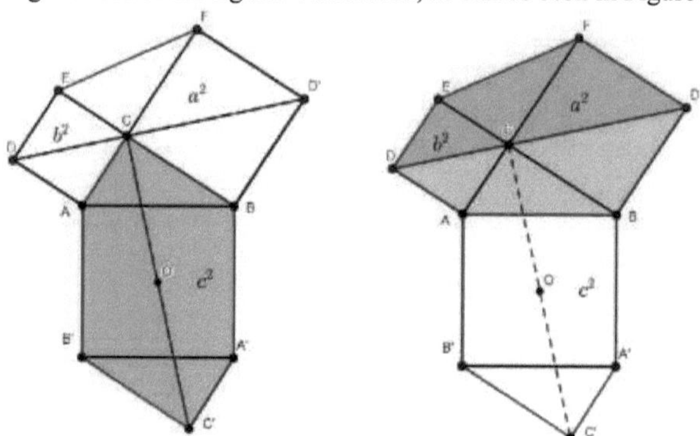

Figure 44. Demonstration of Leonardo Da Vinci with Areas

Source: Own elaboration with GeoGebra software.

Hernandez (2017) rewrites this proof using algebraic language as follows:

We start with a right triangle A *(ABC),* and on each of its sides we draw an external square. Then we draw the right triangle A *(ECF)* which is the symmetric of the triangle A *(ABC)* with respect to the straight line *DD'*, and the triangle A *(Д'C'B')* which is the symmetric of the triangle A *(ABC) with respect* to the centre of the square drawn on its hypotenuse (Hernandez, 2017, p.4).

In this case the verification of Leonardo Da Vinci is an *intellectual proof of visual support* according to the Taxonomy of Proofs.

3.6.2 Pythagoras Theorem according to Anaricio Gopel

Anaricio Gopel was a German who was born in 1812 and died in 1847, and little is known about him, but around 1824, he presents the following proof of the Pythagorean Theorem, see Figure 45.

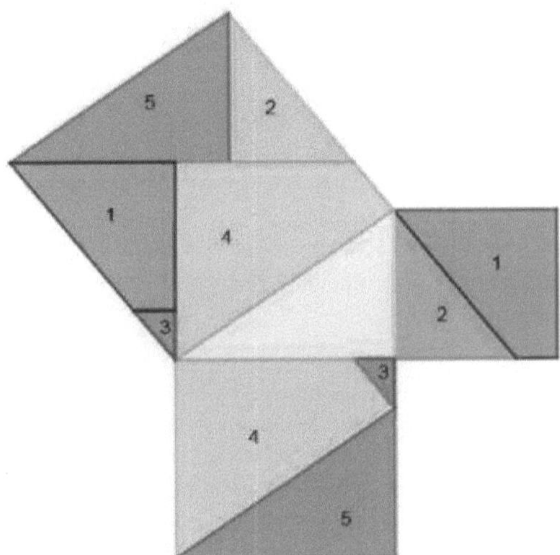

Figure 45. Gopel demonstration

Figure 45 shows a polygonal tessellation of the outer square built on the hypotenuse in five pobgons which, conveniently rearranged, also tessellate the squares built on the legs. For a better understanding, the congruent figures are in the same colour and numbered (Gonzalez Urbareja, 2008).

Gopel performs an *intellectual* test *of visual support* according to the Taxonomy of Tests.

3.7 The Pythagorean Theorem in the Contemporary Age

The Contemporary Age begins with the French Revolution from about 1789 to the present day. In this stage, the Perigal, Garfield and Lamaca tests will be highlighted for this work.

3.7.1 Pythagoras' Theorem according to Perigal

Henry Perigal was born in London on 1 April 1801, died at the age of 97, and was known as Henry Perigal junior, until he was 66, because his father lived to be 100. He was an amateur mathematician, and in 1830 created a proof of the Pythagorean Theorem. His greatest discovery was an elegant cut-and-shift proof, known as the *Dissection Proof of the Pythagorean Theorem,* which he apparently asked to have carved on his tombstone, as seen in Figure 46.

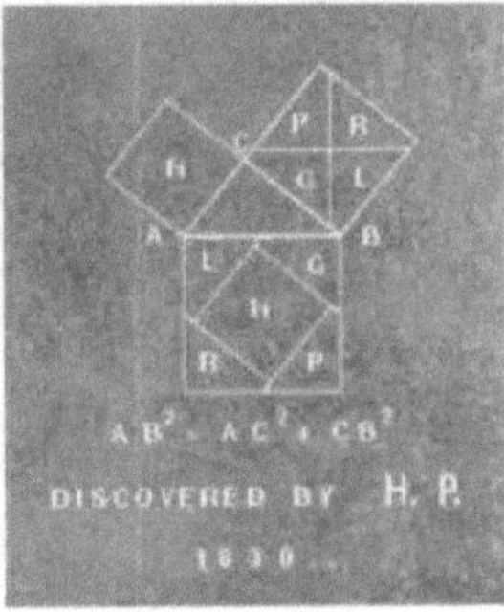

From 1868 to 1897 he was a Fellow of the London Mathematical Society, Treasurer and Fellow of the Royal Meteorological Society for 45 years (1853-until his death in 1898), an elected Fellow of the Royal Astronomical Society in 1850, and a Fellow of the British Astronomical Association from 1890. So it is believed that he was not such an amateur. In his pamphlet *Geometric Dissections and Transpositions* (London: Bell & Sons, 1891), on page 1 of that pamphlet, in the introduction extraichi of *The Messenger of Mathematics* , New Series, No. 19, 1872, Perigal writes:

The present series of articles consists mainly of notes and diagrams taken from my notebooks and sketches made during the last forty years. I have devoted much time and thought to the solution of geometrical theorems and problems by dissections and transpositions: viz. in proving the equality of areas in equivalent rectangles, etc., and investigating how figures might best be dissected, so that their component parts might fit together in any form; and I have often contemplated proving all suitable theorems and problems in Euclid by such dissections and transpositions: to make them self-evident by ocular demonstration. I worked on paper lined all over in little squares, which I found useful to facilitate the dissections; and I have always fancied that the ancient Egyptians and the early geometricians of Greck adopted some of these expedients in their geometrical investigations. So I consider it probable that the property of the right-angled triangle was discovered by similar means; and the solution I found forty years ago was perhaps the same discovered forty centuries ago, and rediscovered by Pythagoras nearly twenty centuries later; and it is not improbable, as was also my case, in the effort to find geometrically a square demonstrably equal in area to a circle by dissection and transposition. "

See page 1 of the booklet *Geometric Dissections and Transpositions*, see Figure 47.

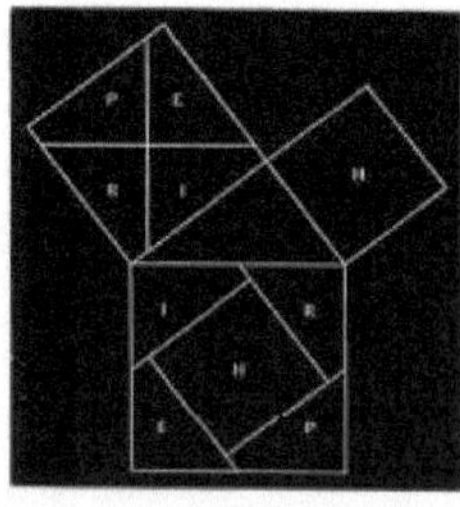

Figure 47. Page 1 of Geometric Dissections and Transpositions, showing Perigal's proof based on the dissection of the Pythagorean Theorem.
Source: http://www.madrimasd.org/blogs/matematicas/files/2019/05/440px-Perigal_1891_p1.jpg

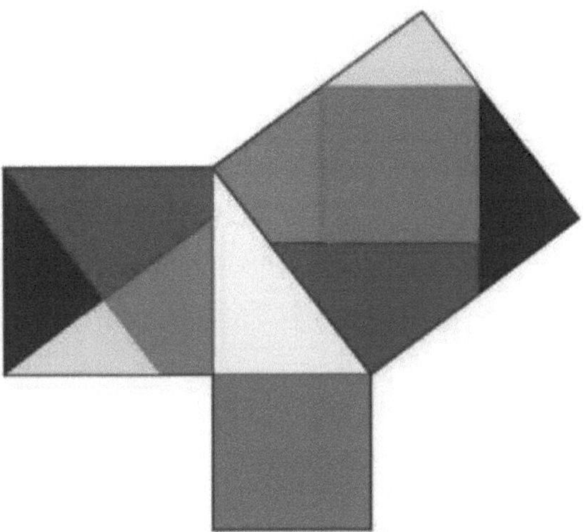

Figure 48. Perigal test

Source: Own elaboration with GeoGebra software based on the text Acevedo (2011).

In the Perigal Test in Figure 48, the outer square on the larger of the legs of the right triangle is tessellated into four parts. While the square outside the smaller cathetus is not divided. Subsequently, the four pieces of the outer square are moved in parallel on the larger cathetus together with the lilac square of the figure, in such a way that they are juxtaposed to the outer square built on the hypotenuse (Acevedo, 2011).

In this case the test is an *intellectual test of visual support*, according to the Taxonomy of Tests developed in section 2.2.

3.7.2 The Pythagorean Theorem according to James A. Garfield

James Abram Garfield lived from 1831 to 1881 and was President of the United States of America. Figure 49 shows Garfield's calling card, which was used at the time as a presentation.

Figure 49. Business card (CDV) or business card with a photograph (from WD Gates & Co.) of James A. Garfield circa 1881 from the author's collection

In 1876 he found and published, in the New England Journal of Education, an original proof of the Pythagorean Theorem shown in the detailed Figure below.

PONS AS/NORUM.

[In a personal interview with Gen. James A. Garfield, Member of Congreae from Ohio, we were shown the following demonstration of the *f>oni asinerwn,* which he had hit open in some mathematical amusements and discussions with other M. C.'s. We do not remember to hare seen it before, and we think it something on which the members of both houses can unite without distinction of party].

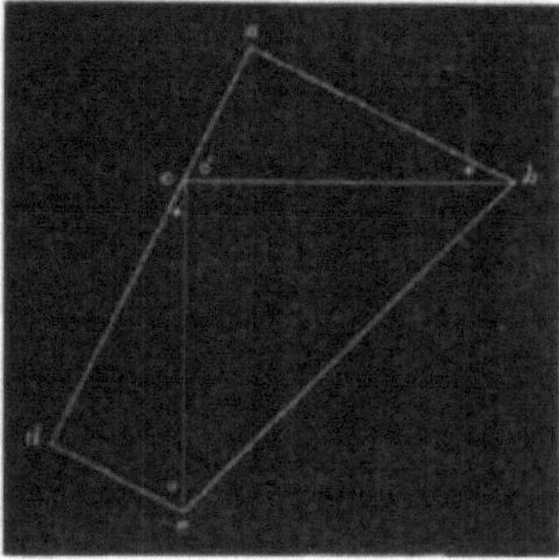

On the hypothenuae *eb of* the right-angled triangle air, draw the half-square cAe. From < let fall the perpendicular *rd,* upon the side *ar* produced.

The triangles *abc* and *dee* are equal; the side *ab^dc,* and the side *ar* e *dr.*

The area of the quadrilateral *adbe* is measured by its base *ad,* multiplied by half the sum *of* its parallel sides *dr* and *ai,* or *adx* which is *at*

But the area of the quadrilateral *adbt* consists of half of the

square of *be* plus the two equal triangles *acb* and *dee* ; or - + +

abXac.-- -|- af X   Nº; or *eb* + *i(ab* X er) -= *ab';*   -|- af X Nº; or *eb* + *i(ab* X er) -= *ab'.*

-|-"t'Sc) -|-ar . *eb ^ab -j-ar . Q. E, D.* [J. A. G.

Figura 49.   Garfield's proof of the Pythagorean Theorem on page 161 of the New-England Journal of Education, 1 April 1876.

Figure 51 is the Garfield proof made in GeoGebra, below is the explanation by Hernandez Cruz (2019) of it.

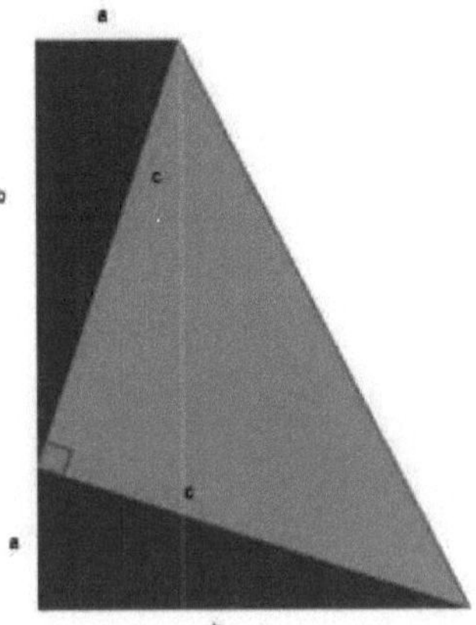

Figure 51. James A. Garfield test

Source: Own elaboration with GeoGebra software based on the text Hernandez, 2019.

According to Hernandez Cruz (2019), Garfield performed the following test. Figure 51 shows how from the right triangle with sides a, b and c, a trapezoid of height (a+b) and bases a and b is constructed.

The trapezium is tessellated into three right triangles: two congruent to the given one, and a third, isosceles of legs c.

The area of the trapezoid is obtained as:

$$Area_{trapecio} = \frac{(a+b)}{2} \cdot (a+b)$$

In turn, the area of the trapezoid can be calculated as the sum of the areas of the three triangles, considering that two of them are congruent, it results:

$$Area = A_{T1} + A_{T2} + A_{T3}$$

$$Area = 2 \cdot \frac{ab}{2} + \frac{c^2}{2}$$

Equating both equations gives

$$\frac{(a+b)}{2} \cdot (a+b) = 2 \cdot \frac{ab}{2} + \frac{c^2}{2}$$

Working both members algebraically, it turns out:

$$\frac{(a+b)^2}{2} = \frac{2ab + c^2}{2}$$

$$(a+b)^2 = 2ab + c^2$$

$$a^2 + b^2 + 2ab = 2ab + c^2$$

$$a^2 + b^2 = c^2$$

The Pythagorean Theorem is thus verified, in this case it is an *Intellectual Proof of formal, geometric and algebraic demonstration,* according to the Taxonomy of Proofs.

3.7.3 Pythagoras' Theorem according to Josep Marfa Lamarca

The Catalan professor Josep Maria Lamarca, in 2006 in the city of Paris, presented in the magazine SUMA the following proof of the Pythagorean Theorem, considering Figure 52:

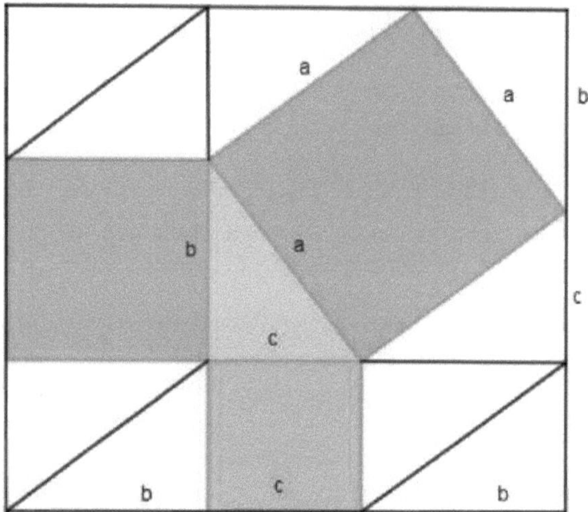

Figure 52. Josep Maria Lamarca's test

Source: Own Elaboration with GeoGebra software based on the text of Lamarca, 2006.

According to the previous figure, given the right triangle with hypotenuse **a** and legs **b** and **c**, the orange squares are built on them. This figure known as the "mill" is inscribed in a square of side (2b + c). Ten right triangles congruent to the initial one and three squares are determined.

Lamarca (2006, p.142) indicates that, when calculating in two different ways, the area of the constructed rectangle, either:

A) by the product of the base (b + c + b) and the height (c + c + b):

$$A = (b + c + b) \cdot (c + c + b) = (2b + c) \cdot (2c + b) =$$

$$= 4bc + 2c^2 + 2b^2 + bc = 2c^2 + 2b^2 + 5bc$$

B) or by the sum of the three square areas and ten triangular areas whose juxtaposition produces the rectangle under consideration:

$$B = a^2 + b^2 + c^2 + 10\,(bc/2)$$

Equating both equations gives

$$2c^2 + 2b^2 + 5bc = a^2 + b^2 + c^2 + 10\,(bc/2)$$

And by simplifying this equality, it turns out:

$$a^2 + b^2 = c^2$$

In this case, the proof corresponds to one of the *Intellectual* type *of formal, geometric and algebraic demonstration,* according to the Taxonomy of Proofs developed in section 2.2.

3.8 Typologies of proofs and demonstrations of the Pythagorean theorem a throughout history

The following Table 2 shows a classification of the different proofs and demonstrations of the Pythagorean Theorem developed throughout history, divided into three stages: the Pre-Hellenic, the Greek and those carried out during the Middle Ages, the Modern and the Contemporary Ages.

Table 2. Typology of proofs and demonstrations of the Pythagorean Theorem in the Pre-Hellenic Age

Pre-Hellenic Stage	Types of tests
Babylon	Pragmatic test of generic model
Ancient Egypt	Pragmatic test of generic model
Former India	Pragmatic test of generic model
Ancient China	Intellectual visual support test

Typology of Proofs and Demonstrations of the Pythagorean Theorem in the Greek Stage

Greek Stage	Types of tests
Pythagoras of Samos Test 1	Intellectual proof of formal and geometrical demonstration
Pythagoras of Samos Test 2	Intellectual proof of formal and geometrical demonstration
Plato's School	Intellectual visual support test
Euclid	Intellectual proof of formal and geometrical demonstration
Pappus	Intellectual proof of formal and geometrical demonstration

Typology of the proofs and demonstrations of the Pythagorean Theorem in the Middle Ages.

Middle Ages	Types of tests
Thabit Ibn Qurra	Intellectual proof of formal and geometrical demonstration
Bhaskara	Intellectual proof of formal and geometrical demonstration

Typology of proofs and demonstrations of the Pythagorean Theorem in the Modern Age.

Modern Age	Types of tests
Leonardo da Vinci	Intellectual Visual Support Test
Gopel	Intellectual Visual Support Test

Table 6. Typology of proofs and demonstrations of the Pythagorean Theorem in the Contemporary Age

Contemporary Age	Types of tests
Perigal	Intellectual visual support test
James A. Garfield	Intellectual Test Formal, geometrical and algebraic demonstration
Josep Maria Lamarca	Intellectual Test Formal, geometrical and algebraic demonstration

Chapter 4

Proposed teaching sequences

"Knowing mathematics is not just about learning definitions and theorems, to recognise the opportunity to use and apply them, we know well that doing mathematics implies that one is engaged in solving problems, but sometimes it is forgotten that solving a problem is only part of the job; finding good questions is as important as finding the solution to them".

Brousseau, 1994

4.1 Mathematical demonstration from a didactic point of view

This chapter will refer to mathematical demonstration from a didactic point of view, analysing the ways of understanding and expressing mathematical knowledge.

Throughout history, countless controversies have arisen among mathematicians about the acceptance or not of certain characteristics of mathematical work, the role of demonstrations and definitions, the level of rigour required and the different approaches that can be taken to mathematical knowledge (Crespo et al., 2010, p. 14).

Several authors agree that mathematical knowledge is based on understanding and expression, which can be acquired in two different ways, *directly* or *reflexively*. Both modes of knowledge are complementary and are combined in the process by which mathematical objects, their relationships and the way in which it is possible to operate or interact with them are described (Crespo et al., 2010, pp. 14-15).

Table 7. Modes of understanding and expressing mathematical knowledge

Direct form	Reflective form
• Creative- Anabtico	
• Subjective - Reflective	

The teaching of mathematics requires inductive and deductive reasoning, and this does not refer to the use of an axiomatic system at an early age, but rather to begin to encourage young students to intuit, to elaborate hypotheses, to carry out proofs without the need to formalise. It is only around the age of nine or ten that the capacity for abstraction necessary to begin to internalise formal thinking begins to form in the human being (Santalo, 1966). This level of formal and axiomatic analysis begins to develop in late adolescence. Reasoning provides a powerful way to develop knowledge and logic, and also provides students with the critical ability to detect inconsistencies that will allow them to develop various stages of learning (Crespo et al., 2010).

"Reasoning and demonstration should be a consistent part of the mathematical experience throughout schooling. Mathematical reasoning is a habit of mind and, like all habits, must be developed through consistent use in many contexts" .... (NCTM, 2000, 59).

The aforementioned highlights the importance of thinking about the role of demonstration in secondary school. The authors Ibanes and Ortega (1997), wrote about learning to demonstrate, considering two types of activities: the first aimed at *understanding* the demonstration and the second aimed at *doing* demonstrations. To understand the demonstration "Generally, demonstrations are done in front of the students and then they are asked to do the same" (Report of the First Cycle Group, Jornadas Nacionales APMEP, 1979). But Ibanes and Ortega (1997) propose something else: to understand a demonstration, the teacher must show one, and the student must understand it. To achieve this, it is necessary to understand it globally, which implies possessing a scheme of proof, which is particular to each person and consists of what constitutes discovery and persuasion for them.

Figure 53 shows the demonstration learning and understanding steps proposed by these authors.

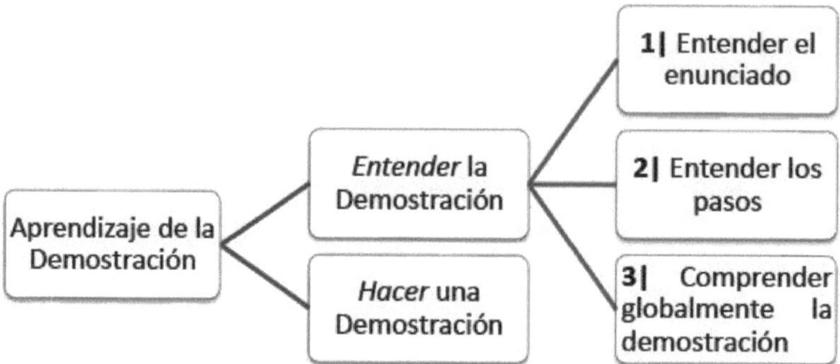

Figure 53. Learning from demonstration

Once the student understands the demonstration, he or she will be able to do one on his or her own. What better resource than to use the History of Mathematics to use demonstrations that help students to understand them, and thus to learn them.

4.2 Taxonomy of tasks to address geometric notions

For the authors Garda and Lopez (2008) there are three types of tasks that are performed in the classroom when studying geometric notions: *conceptualisation, investigation* and *demonstration*, which are expected to develop students' geometric reasoning. They can occur simultaneously in the solution of the problem situations they are presented with, and it is often very difficult to differentiate between them. An investigation task can lead to the construction of a geometrical concept and at the same time lead to the argumentation of the results, like a demonstration task.

These authors define them as follows:

• **Conceptualisation tasks**: these refer to the construction of concepts and geometrical relations. It is important to clarify that it is not about defining geometrical objects but conceptualising them.

• **Investigation tasks**: these are tasks in which the student investigates the characteristics, properties and relationships between geometrical objects with the purpose of giving them meanings. It is probably in this type of task that the problem-solving approach to geometry teaching is best appreciated.

• **Demonstration tasks**: these tend to develop pupils' ability to elaborate conjectures or procedures for solving a problem which they will then have to explain, prove or demonstrate on the basis of arguments that can convince others of their veracity. It is in this type of activities where the socialisation of geometric knowledge can be appreciated, since the problem-solving approach conceives knowledge as a social construction. Demonstration tasks are essential in Geometry and should be present in the classroom interaction; the construction of logical arguments is a skill that is an essential part of geometric culture and it is desirable that all students develop it (Garcia and Lopez, 2008, p. 42).

The tasks required to address the geometric notions proposed by Garda and Lopez (2008) are shown in Figure 54, and will be taken into account when developing the objectives and activities of the teaching sequences proposed for the particular case of the Pythagorean Theorem.

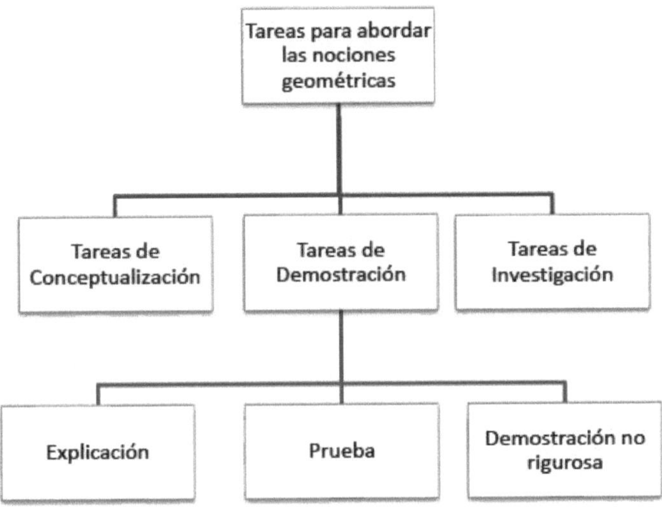

Figure 54. Taxonomy of demonstration tasks to address geometric notions in schooling.

In addition, for the elaboration of the didactic sequences, the following functions will be taken into account
of Michael de Villiers' (1993) demonstration, which are summarised as follows:

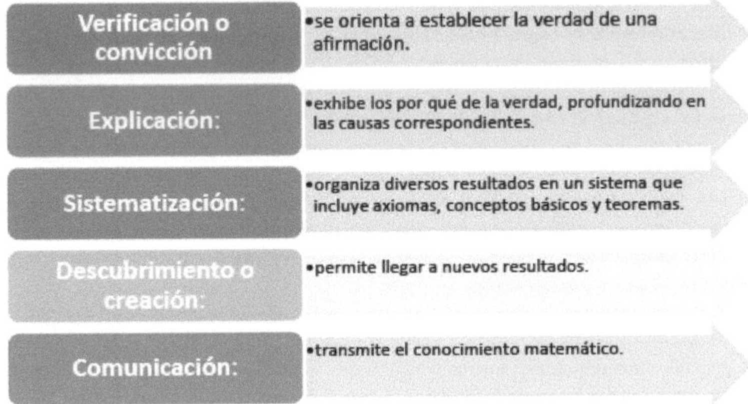

Figure 55. Functions of the Demonstration

4.2.1 Demonstration tasks in Geometry class

For Hanna and Jahnke (1996; p. 877) "demonstration is an essential feature of mathematics and as such should be a key component of mathematics education". This raises the question ^what role do demonstration tasks play in mathematics classrooms?

What is known as "modern mathematics" is a movement that during the sixties incorporated into the educational system several reforms related to the teaching of mathematics, bridging the gap between school and scientific knowledge. In our country the incorporations took place in the sixties and seventies. The deductive procedure of space geometry was abandoned because it was considered an elaboration of plane geometry, and coordinate geometry was introduced to the point where the theorems could be simply demonstrated by algebraic methods. According to Fehr (1972) quoted in Cuadernos de Investigacion y Formacion en Educacion Matematica, "Every future citizen should experience the study of an axiomatic structure to know what mathematicians do to control all discoveries" (p.57). This sentence marks the importance of axiomatisation in the educational system of the time.

From the 1990s to the present day, there have been many debates about the definition of the tasks of demonstration. For this reason, and as mentioned in the theoretical framework, demonstration is a complex subject, linked to others such as explanation, verification, argumentation, proof, etc. The authors Torregrosa Girones, Delicado, Martinez and Ciscar (2010) focus on characterising the development of the cognitive processes that students show when solving geometry problems and mention some of the demonstration tasks:

* *explain* why a mathematical result is true,
* *communicating and conveying* mathematical relations and properties,
* *develop logical and abstract thinking*,
* establishing a *link between mathematics and reality*,
* *systematise,* organising the results in a deductive system of axioms and theorems,
* *discovering and constructing* mathematical knowledge.

In other words, when developing teaching sequences, the aim should be for students to fulfil the above-mentioned tasks. But this should be circumscribed to Argentine legislation, which has a National Education System comprising four levels: Early Education, Primary Education, Secondary Education and Higher Education, and eight modalities. This organisation was established in the National Education Law N° 26.206 sanctioned in 2006, but previously the National Education System contemplated an initial level (kindergarten), the basic general education (EGB) of three cycles of three years each, which replaced primary school, and which extended compulsory education until the age of 15. Finally, the Polimodal, a level lasting three years. One of the documents of the time stated, with regard to demonstration:

Throughout primary school, the contrasting of concepts and relationships, the search for regularities in a set of data (facts, forms, numbers, algebraic expressions, graphs, etc.) and the formulation of generalisations on the basis of observation, experience or intuition, all point to the formation of inductive reasoning, not "mathematical induction", which is a method of demonstration intrinsically included in the concept of natural number. Mathematics uses induction as a starting point, but the truth of its propositions is proved by **deduction**.

Deductive reasoning proves the formal truth of its conclusions as a "necessary" derivation of its premises.

Proving a generalisation requires deduction which makes it independent of experience and universal. Deductive reasoning is not necessarily linked to a formal presentation of it, and at this level such a presentation is not necessarily necessary, but it is interesting for students to be able to use and differentiate between the different forms of verification.

(Ministry of Culture and Education, 1995, p. 87).

In this document, on the one hand, it can be observed that the modes of proof are called verifications and that, as mentioned above, deduction is used, but not from a formal point of view, since intermediate level students would not be able to reach such a level of abstraction.

The following is an analysis of the treatment of demonstration in Argentina's current legislation:

As for deductive reasoning, this way of thinking will be approached in all the axes, but it will be more

56

heavily involved in the axes "Numbers and operations" and "Geometry and magnitudes".

For this approach to be possible, it will be necessary for pupils to recognise the value of deduction as a means of verifying the validity of a mathematical statement. To this end, situations will have to be proposed which cannot be resolved through measurement, perception or exemplification, and which place the pupils before the need to produce arguments which demonstrate their validity without resorting to empirical verification. In this way, deduction will not come into play as a requirement of the teacher but as a necessity of the situation itself.

It should be understood that the possibility for pupils to make demonstrations also involves a slow and complex process, so that more or less precise or formal arguments - from a mathematical point of view - should be accepted so that they gradually approach more formal deductive arguments with the help of the teacher. In the first years of secondary school, the differences between a demonstration, a justification, an explanation or a convincing argumentation will be much less important than their similarities.

Mathematical formality should not be the starting point of the deductive task but the goal (Direction General de Cultura y Education, 2007, pp. 297-298).

It will not be relevant for the deduction of new properties if the pupils are not yet familiar with symbolic formality. Forcing them to work in this way could prevent them from achieving the desired objective. Teachers should bear in mind that they are proposing a gradual approach to a geometry centred on demonstration which will continue to be consolidated in the following years of secondary school.

The students will be able to elaborate, at the beginning, arguments in colloquial language. Throughout the year, the teacher will encourage the achievement of increasingly clearer and more formal levels of argumentation from a mathematical point of view.

(Direccion General de Cultura y Educacion, 2007, p. 313).

In the preceding pages of the Curriculum Design, it can be seen that the deductive method is introduced, but as a necessity on the part of the students and as something imposed by the teacher. As demonstration is not considered from a formal point of view, it is not necessary to distinguish between the terms demonstration, justification, explanation or argumentation, because it is not considered that the student can use colloquial language to *"demonstrate"* without resorting to symbols and formalisms. However, in this work it is considered necessary to distinguish the terms mentioned above, as was explained in Chapter 2, but with the understanding that the developmental horizon of secondary school students will not allow them to reach the levels of abstraction necessary to be able to carry out formal demonstrations.

The following excerpt from the Curriculum Design for Secondary Education points out that failure in Mathematics is sometimes due to the mistaken belief that the mathematical endeavour is only for the few:

Although school mathematics differs from scientific work, the style and characteristics of the work of the mathematical community can and should be experienced in the classroom. In this way, students will see the discipline as a possible task for everyone, as defined in the Basic Cycle of Secondary School.

In general, school mathematics is characterised by a profusion of abstract definitions, mechanistic procedures, univocal and finished developments, and formal demonstrations together with a hasty use of symbolism. This contributes to the belief that people who are unable to assimilate the knowledge associated with it in a systematic way, in the order and quantity in which it is presented, fail because of a lack of ability in the subject.

This deterministic and elitist conception contrasts with the proposal of the present Curricular Design, which considers the discipline as part of the culture and values the students as makers of it.

(Direccion General de Cultura y Educacion de la provincia de Buenos Aires, 2010, p. 9).

On the other hand, related to Pythagoras' Theorem, in the Priority Learning Nuclei (NAP) for Secondary Education in Argentina (p. 17), in relation to Geometry and Measurement, it is mentioned about the analysis and construction of figures, arguing on the basis of properties, in problematic situations that require:

- That students should be able to analyse the relationships between sides of triangles whose measures are Pythagorean triads and interpret some proofs of the Pythagorean Theorem based on equivalence of areas.

At provincial level, in Mendoza, it is established in relation to the Provincial Curricular Designs, regarding the Axis: Geometry and Measurement, that students can interpret some demonstrations of the Pythagorean Theorem to be applied in different situations. In accordance with this, the General Directorate of Culture and Education (2007), mentions with regard to the Pythagorean Theorem: *"Check with the help of the teacher the validity of the Pythagorean Theorem"* (p.311). In the same document but on page 316 it is quoted:

In this axis (Geometry and magnitudes) the properties of right triangles will also be studied, in particular the Pythagorean Theorem, based on problems that give it meaning.

You can find several "demonstrations" of the theorem to be analysed, although not all are of the same level of difficulty, it is interesting, if you have the means, to take the opportunity to visit it.

(Direccion General de Cultura y Educacion, 2007, p. 316).

In accordance with the curricular designs, it is proposed to approach demonstration in the mathematics class and in particular in relation to the Pythagorean Theorem, through the analysis and interpretation of some demonstrations with different levels of difficulty, mainly based on equivalences of areas, trying to check their validity. Understanding that the work of the mathematical demonstration task itself will allow the student to develop skills of argumentation, communication, logical and abstract thinking, and systematisation, among others.

4.3 Incorporation of ICTs in the sequences

The role of Information and Communication Technologies (ICT) in the current teaching and learning process should not be overlooked. For a long time it was considered that only the constructions with ungraded ruler and compasses were valid, then man was creating other tools that allowed him to graph, test and work in Mathematics incorporating the new technologies. The incorporation of ICT provoked Florencia Veronica Aspera great debates about the improvement or not in the teaching and learning process, some of which are still present today.

Here is an example where you can read a teacher's testimony about the importance of dynamic demonstration in the development of students' knowledge.

For our students, dynamic demonstration often allows them to be convinced more effectively and to analyse formulas and hypotheses deductively, which is why, for them, dynamic demonstration constitutes a fundamental pillar on which to develop their knowledge. (First forum. Teacher AnDi) (emphasis added) (Torregrosa et al, 2010, p.394).

ICT is considered to improve the teaching and learning process because it favours among other things the possibility of manipulating mathematical objects and analysing relationships between them easily and quickly. Thus, in the sequences of this work, software such as GeoGebra will be used because "the graphic and manipulative characteristics of the dynamic *software* allow experimentation, conjecture and verification, and this makes it a powerful tool that will allow the student to generate propositions through the identification of particular cases" (Torregrosa et al, 2010, p.396).

58

4.4 **Proposed teaching sequences**

According to Rodriguez (2012), a didactic sequence is a collection of instructions or activities selected with certain objectives, considering the students' prior knowledge, the context and the characteristics of teaching with which they have been working and the classroom management. A didactic sequence has the purpose of ordering and guiding the teaching process promoted by an educator. In general, this set of activities is indicated within a systematic educational process linked to a given object of study. In addition, the proposed activities must share a common thread that enables students to develop their learning in an articulated and coherent way.

The didactic sequences elaborated in this work are designed for a group of students who have the previous knowledge detailed in them, who have access to the materials and resources necessary to carry out the proposals, as well as considering that they have been working in a similar way to the one proposed. This is to emphasise that teachers who read these proposals and want to use them in the classroom should take these considerations into account, and above all adapt them to their students in the way they consider most appropriate. It is also worth noting that they are designed under a constructivist paradigm, understanding that "constructivist theories are, above all, epistemological theories; that is, they are theories that provide us with an explanation of how knowledge is produced, and what are the conditions for this production to take place" (Waldegg, 1998, p.16). Under this paradigm, the teacher plays the role of mediator between knowledge and the student; classes are no longer merely expository, but the student learns by doing. This makes it possible to develop cognitive competences that favour the construction of knowledge, in this case mathematical knowledge, promoting responsible and collaborative participation. In this way, the necessary competences are developed by using activities that require the use of concrete materials such as the tangram, pencil and paper, and the GeoGebra software. This software was chosen because it is free, free of charge, can be used online, can be downloaded to the computer or mobile phone, and has user manuals and construction protocols.

In this document, access to the historical construction of knowledge, in this case the Pythagorean Theorem, is considered fundamental, taking as a basis the historical proofs and demonstrations of this theorem, in order to elaborate didactic sequences that allow the basic competences mentioned above to be deployed.

During the learning process, the student, through active, conscious and responsible participation, constructs his or her own learning, giving meaning and re-signifying his or her schemes of knowledge in order to achieve intellectual autonomy. Baroody and Dowker (2003) distinguish four teaching methods centred on the mathematical activity being developed, summarised by de Castro (2007, pp. 64-65):

The *skills approach* focuses on the memorisation of basic skills through repetition. This approach is based on the assumption that mathematical knowledge is a collection of rules, formulas and procedures. Learners are seen as empty vessels, unable to grasp most mathematical knowledge. The most efficient way of teaching will consist of direct teaching of procedures, followed by a great deal of practice. No attention is paid to the understanding of procedures. Teaching and practice tend to make little reference to context and tend to be highly symbolic (abstract). Activities do not have a clear meaning (why) for the learners, are often not based on their interests, do not involve genuine mathematical activity, and are not meaningful. However, learners can become very skilled at executing procedures, being very fast and making few mistakes.

The *conceptual approach* focuses on learning procedures with understanding. Mathematics is seen as a network of concepts and procedures. Children are considered capable of doing mathematics as long as they are taught how the procedures work. The aim of this approach is for children to learn the rules, formulas and procedures in a meaningful and comprehensible way. The symbolic procedures are represented by concrete models, using drawings or manipulative materials. Although the activities are sometimes decontextualised and their meaning (why they are done) is not clear, there is an effort to promote meaningful learning.

The *problem-solving approach* is radically opposed to the skills approach. It focuses on the development of mathematical thinking through reasoning and problem solving. Mathematics is seen as a way of thinking, a process of investigation, or as the search for regularities in order to solve problems. Children are seen as possessing immature thinking and incomplete knowledge, but they are endowed with great natural curiosity and are able to actively construct their own knowledge and understanding of mathematics. The main objective of the teaching is to introduce the beginner to mathematical activity through solving real problems for children. The teacher acts as a partner in the process of investigation without directing this process. In this approach, the learning of procedures is secondary to the development of mathematical thinking.

The *investigative approach* is a mixture of the conceptual and problem-solving approaches. Mathematics is seen simultaneously as a network of concepts and procedures and as a process of inquiry. Children are seen as capable of actively constructing their knowledge, a construction that is mediated and guided by the teacher through pre-planned activity proposals, but also through investigative experiences that emerge during the learning process. The aim is to learn rules, procedures and formulas in a meaningful way, but also to acquire reasoning, representation, communication and problem-solving skills.

We must not lose sight of the fact that the basic competences of mathematical work are *reasoning, communication* and *problem solving*. In this chapter, two didactic sequences are developed, which aim to achieve these skills, under the investigative approach mentioned in the previous paragraph. It is intended that through the proposed activities, mathematical concepts and procedures are interwoven, all of them under the historical construction of the Pythagorean Theorem.

Each of the didactic sequences and their respective activities are designed with an increasing degree of difficulty, starting from more concrete activities, to finally reach a higher degree of abstraction. It can be seen that, in turn, the activities in the second proposal are more complex than those found in the first.

In addition, the slogans were selected by analysing their mathematical potential, understanding that the latter is related to the *possibilities of* exploration that the slogan enables or does not enable and with the *possibilities of arguing about the validity of the resolution or its response* (Rodriguez, 2012, p. 3). Thus, Mabel Rodriguez (2012) establishes that the mathematical potential (MP) of a slogan can be assessed qualitatively as poor MP, rich MP and the intermediate possibilities that exist. Considering the first case as those slogans that do not admit exploration and do not require argumentation, while the second when they do admit it and require it.

Again, it is up to those who want to implement such activities in their practice to make their own assessment of the mathematical potential of these activities and above all to adapt them to their realities and their students.

In conjunction, some activities include Information and Communication Technologies (ICT) therefore the TPACK (Technological Pedagogical Content Knowledge) model was taken into account. It is a conceptual theoretical framework that identifies some of the essential qualities of knowledge that teachers need to build in order to integrate new technologies into teaching.

It was developed by Mishra and Koehler in 2006. Figure 56 visualises the relationship between the disciplinary, pedagogical and technological sources of knowledge, as well as the knowledge generated from the interactions between them.

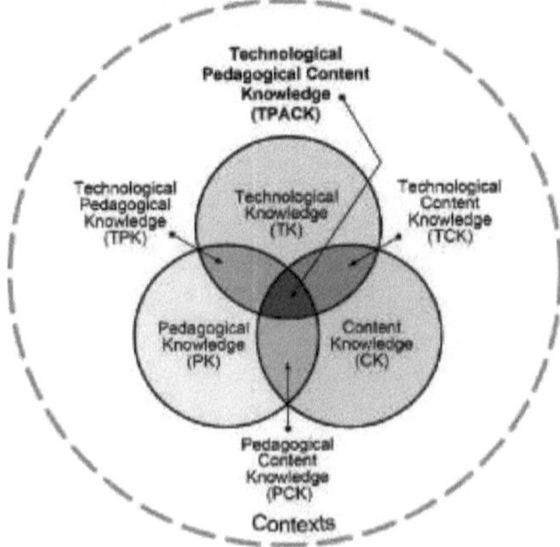

Figure 56: Technological pedagogical and disciplinary knowledge. The three circles - discipline, pedagogy and technology - overlap and generate four new forms of interrelated content.
Source: http://www.tpack.org

The TPACK model was applied to sequence planning, because this model is complex. In this sense, the following were considered:

- *Curricular decisions*: for the choice of the historical proof of the Pythagorean Theorem, for drafting the aims and objectives, and finally for determining the necessary prior knowledge.
- *Pedagogical decisions*: to develop the activities, to determine the roles of the teacher and the learner and to obtain assessment strategies.
- *Technological Decisions*: to search for and choose the ICT resources to be used.

As for the curricular decisions to choose the contents linked to the proofs of the Pythagorean Theorem and prior knowledge, the contents of the NAPs for Secondary Education were taken into account, formulating the following purposes:

- *To present students with situations that allow them to construct knowledge about the Pythagorean Theorem and its applications.*
- *Promote collaborative work, discussion and exchange between peers, the students' autonomy and the role of the teacher as a guide and facilitator of the work.*
- *To recognise the contributions of the History of Mathematics in the construction of knowledge.*
- *Promote the appreciation and use of technological resources for the exploration and formulation of conjectures, for problem solving and for the control of results.*

The following elements are considered in each of the teaching sequences:

- Objectives
- Learning and Content
- Activities
- Resources
- Evaluation

Assessment is considered in its different dimensions (diagnostic, formative and summative) in order to provide students with real possibilities to reflect on their own processes, achievements and difficulties during the teaching and learning process and to verify whether and to what extent the expected learning has been achieved. The evaluation does not only refer to content, but also aims at assessing the competences of the mathematical task. In both didactic proposals, the first activity is related to the *diagnostic assessment,* and the other activities are designed in a spiral fashion where the content is gradually built up and the competences mentioned above are developed. When implementing the activities with the students, interventions, modifications and adjustments will arise, thus developing the *formative evaluation* of the learning process. Finally, the *summative evaluation,* which has the function of assessing whether the objectives and the pedagogical intentionality of the proposal were met, evaluating the entire teaching-learning process. As far as the evaluation instruments are concerned, all the activities of the sequences must be considered as instruments of the tasks, the teacher can record the development of the students in all of them, for example, with an evaluation rubric.

4.4.1 Sequence 1

Level: Lower Secondary Education

Suggested courses: 1st, 2nd and 3rd year

Area: Mathematics

Curricular focus:

Geometry and Measurement

Objectives:

- Recognise, name and classify triangles by their shape and properties.
- Construct plane figures, applying properties and using concrete materials.
- Interpret some proofs and demonstrations of the Pythagorean Theorem based on equivalence of areas.
- To work out, in an algebraic way, proofs and demonstrations of the Pythagorean Theorem using formulas for areas of plane figures.
- Analyse the relationships between sides of triangles whose measures are Pythagorean triads.
- Apply the Pythagorean theorem to problem solving.

Learning and content:

- Construction of plane figures.
- Classification of triangles according to their shape and properties.
- Selection and interpretation of different proofs of the Pythagorean Theorem throughout history.
- Analysis of the Pythagorean triads.

Approach activity:

Group Activity

Resources:

J Triangles made in cardboard (rectangles, acuteangles, obtusangles, equilaterals, isosceles and scalenes).

The pieces are distributed and the following instructions are given:

Observe the different triangles provided, and identify their constituent elements (sides, vertices, angles), using ruler, square, etc., if necessary.

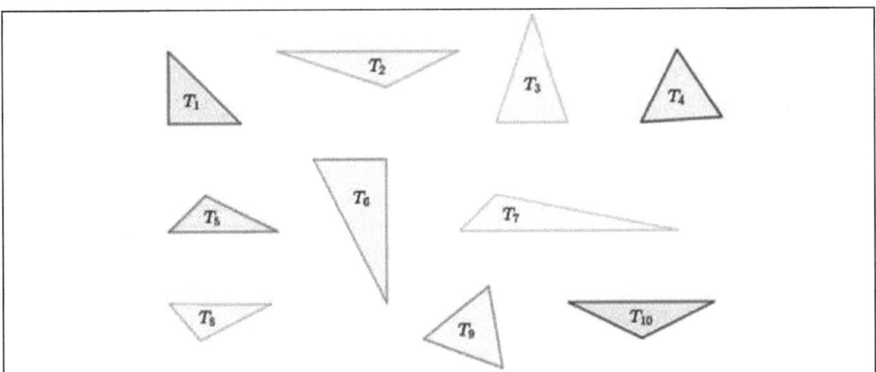

Classify the figures according to their sides and finally according to their angles. Complete the following table:

Classification of triangles					
According to their sides			**According to their angulos**		
Isosceles	Equilatero	Scalene	Acutangulo	Obtusangulo	Rectangle

Knowledge construction activity:

Group Activity

Resources:

J 2 sets of two-coloured Tangrams.

J 1 large rectangular triangle with the dimensions of the Tangram in a different colour.

The Tangram is a very old popular game, coming from China, and it is made up of seven pieces, called "tans":

- A square
- Two large right-angled triangles
- A medium rectangular triangle
- Two small right-angled triangles
- A parallelogram

The teacher facilitates the games in wood or eva rubber.

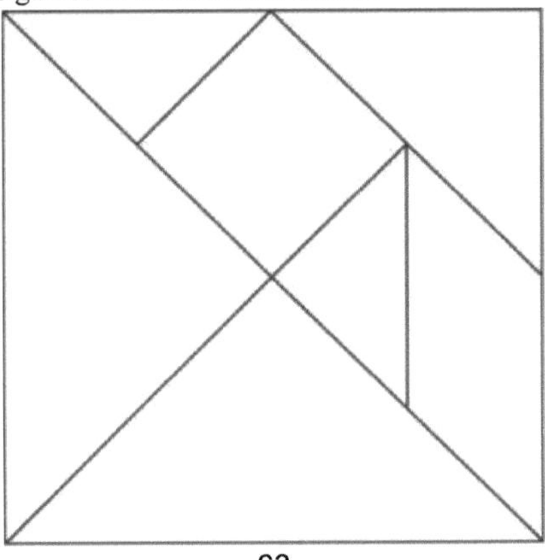

Each group is given two sets of tangrams and a rectangular triangle of a different colour, congruent to the largest triangle in the tangram. The following instructions are given:

1) On the coloured rectangular triangle you must build three squares with the tangram pieces on its sides, using all the pieces (without overlapping them or leaving spaces), the length of the sides of which coincide with those of the triangle.

2) Answer:

a) Compare the areas of the larger square with the sum of the areas of the smaller squares(,are they larger, smaller or equal?

b) How can you justify the above answer?

c) With what type of triangle do we start such a construction?

The student is expected to raise the following:

The teacher takes advantage of the situation and justifies it. Thus, with the students, he formalises the Pythagorean Theorem and, through the superposition of pieces, they justify the theorem.

In any right triangle, the area of the square whose side is the hypotenuse is equal to the sum of the areas of the squares whose sides are each of the legs.

Knowledge construction activity:

Individual Activity

Resources:

J Photocopy with the story and the figure of the Garfield verification

The photocopy shown below is handed out and the following instructions are given:

In 1876 Garfield, an American who later became president, carried out a proof of the Pythagorean theorem. It consisted of a trapezoid like the one in the figure which has been tessellated by three right triangles, two of them congruent (T1 and T2) and T3 being isosceles.

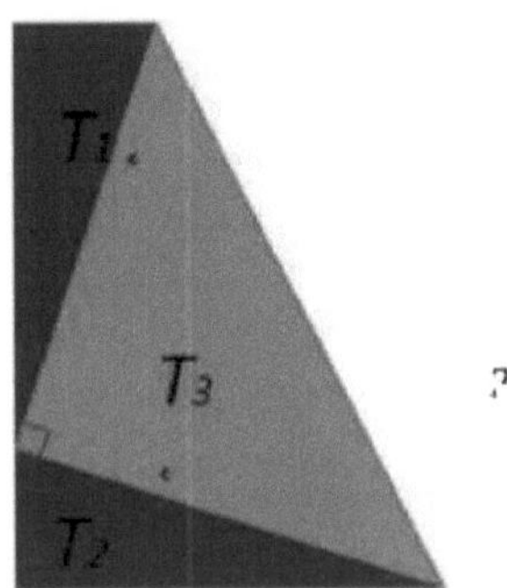

b

Answer the following questions:

a)   State the formula for the area of the trapezium as the sum of the areas of the triangles.

$$^{\wedge}trapezoid = {}^{\wedge}T_1 + {}^{\wedge}T_2 + {}^{\wedge}T_3$$

b)   Work algebraically both members of the equality ^What do you observe? What have you obtained?

The student is expected to raise the following:

a)

$$A_{trapecio} = A_{T_1} + A_{T_2} + A_{T_3}$$

$$\frac{(a+b)}{2}.(a+b) = 2.\frac{ab}{2} + \frac{c^2}{2}$$

b)

Working algebraically both members of the above equation:

$$\frac{(a+b)^2}{2} = \frac{2ab+c^2}{2}$$

$$(a+b)^2 = 2ab + c^2$$

$$a^2 + b^2 + 2ab = 2ab + c^2$$

$$a^2 + b^2 = c^2$$

Implementation activity:
Individual Activity
Resources:
J Photocopy
J Sheet and pencil
J Calculator (if required)

The photocopy shown below is handed out and the following instructions are given:
The triangle with sides 3, 4 and 5 is rectangular, where 3 and 4 are the measures of the legs and 5 is the measure of the hypotenuse. This triangle has the property of having by measure of its three sides whole numbers. It is also known as the sacred Egyptian triangle, and it was used to form right angles that were later used for example in the construction of pyramids and monuments, since they could duplicate, triple and other measures. Therefore (3,4, 5) is known as the Pythagorean tern and applies to right triangles.

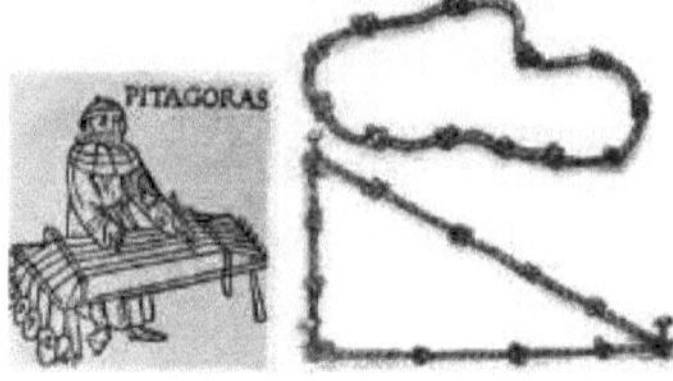

Identify which of the following triads correspond to right triangles by applying the Pythagorean Theorem, and justify your choices:

1.   (3,4,5)
2.   (5, 6, 7)
3.   (6,8,10)
4.   (5,12,13)
5.   (12,35,36)

Implementation activity:

Individual Activity

Resources:

J Photocopy

J Sheet and pencil

J Calculator (if required)

The photocopy shown below is handed out and the following instructions are given:

In China, the Pythagorean Theorem is known as Kon Kun or Gougu. There are two classical Chinese treatises of mathematical content where geometrical aspects linked to the Pythagorean Theorem are related, they are the Chou Pei Suan Ching (300 B.C.) and the Chui Chang Suang Shu (250 B.C.). The latter contains 246 problems of which 24 concern right-angled triangles.

Solve the following problem shown in the figure:

"There is a bamboo ten feet high, which has broken in such a way that its upper end rests on the ground at a distance of three feet from the base. You are asked to calculate at what height the break has occurred".

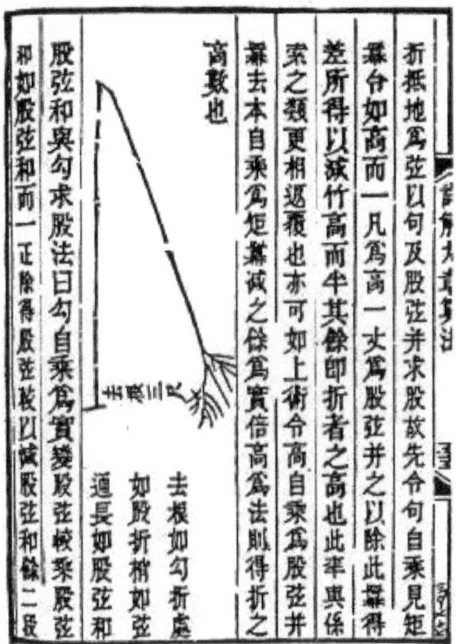

Evaluation Rubric

The proposed rubric was elaborated on the basis of section 4.1. and the Genially tool was used in its construction, which has a free version without the need for a trial period.

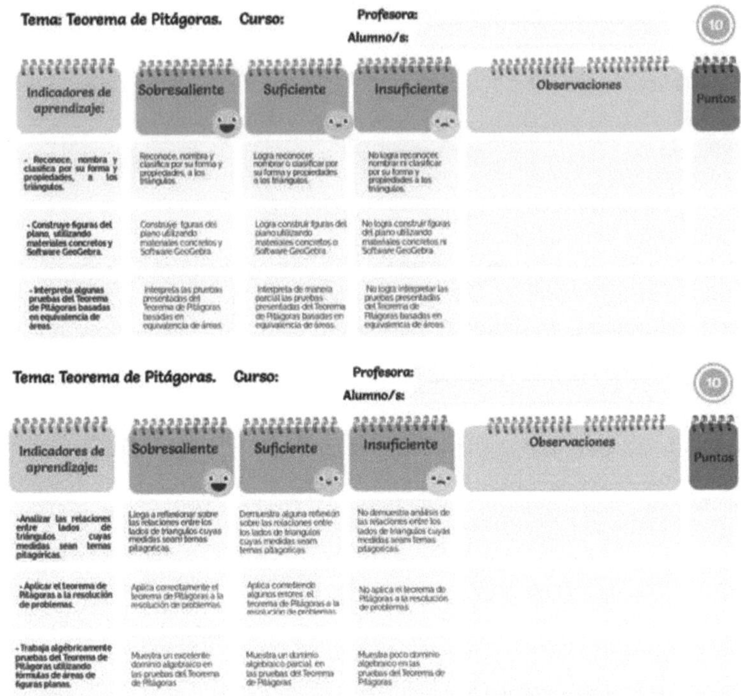

Rubric available at https://view.genial.ly/5f73e78134baa70d0752059f/horizontal infographic-review-rubrica-tesina-secuencia-ndegreel

4.4.2 **Teaching sequence 2**

Level: Lower Secondary Education

Suggested courses: 1st, 2nd and 3rd year

Area: Mathematics

Curricular focus:

Geometry and Measurement

Objectives:

- Construct plane figures, applying properties and using concrete materials and GeoGebra software.

- To recognise and classify triangles by their shape and properties.

- Interpret some proofs and demonstrations of the Pythagorean Theorem based on equivalence of areas.

- To work out, in an algorithmic way, proofs and demonstrations of the Pythagorean Theorem using formulas for the areas of plane figures.

- Analyse the relationships between sides of triangles whose measures are Pythagorean triads.

Learning and content:

- Construction of plane figures.

- Classification of triangles according to their shape and properties.

- Selection and interpretation of different proofs of the Pythagorean Theorem throughout history.

- Analysis of Pythagorean triads.

Approach activity:

Individual Activity

Resources:

J Sheet and pencil

J Geometric instruments

The following instructions are given:

1) Reproduce, using geometry tools, a figure with 3 sides and a right angle.

2) Classify the triangle drawn, according to its sides.

3) Observe and compare the different right triangles drawn by your partners(,is it possible to construct an equilateral right triangle? ¿Why?

The teacher reviews the classification of triangles and recalls the names given to the sides of a right triangle (legs and hypotenuse), as well as the relationship between the measurements of the sides of a triangle and the properties of the interior angles of the triangle.

Knowledge construction activity:

Individual Activity

Resources:

J Photocopies

J Puzzle pieces of Perigal made of wood or eva rubber.

The photocopy and the pieces of the puzzle shown below are handed out and the following instructions are given:

Henry Perigal was born in London on 1 April 1801, he died at the age of 97. Perigal's 6-piece jigsaw puzzle contains:

J a right-angled triangle,

J a square

J four congruent trapezoids, each with two right angles.

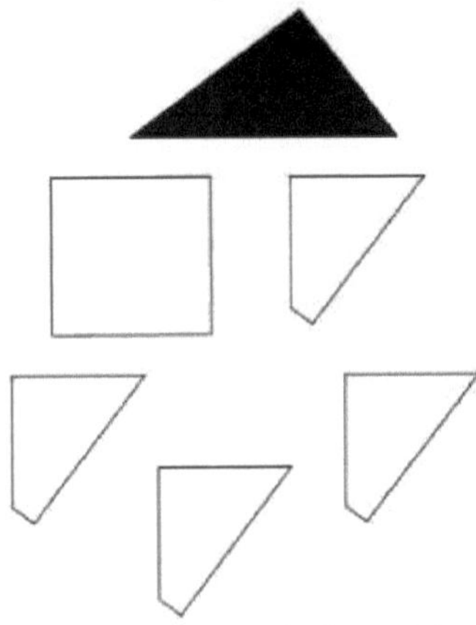

1)   Use the given pieces to form two squares on the legs of the triangle.

2)   Then rearrange the pieces so that they form a square whose side coincides with the hypotenuse.

3)   Answer:

a)   Compare the area of the larger square with the sum of the areas of the smaller squares(, are they larger, smaller or equal?

b)   How can the above answer be justified?

c)   With what type of triangle do we start such a construction?

The student is expected to raise the following:

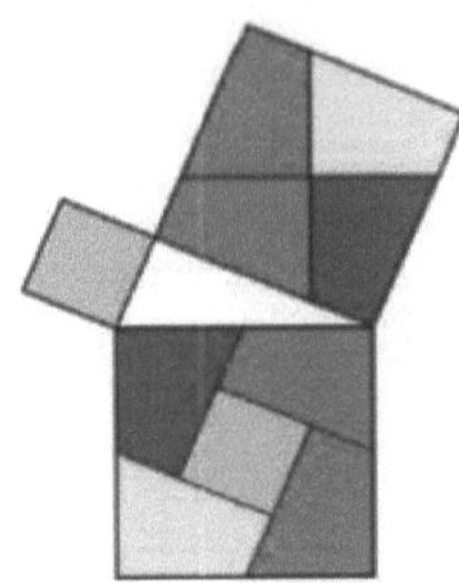

The area of the larger square is equal to the sum of the areas of the smaller squares. This in conclusion derives from the observation that they use to construct the two smaller squares the same pieces used in the construction of the square whose side is the hypotenuse of the right triangle with which we started the construction.

Knowledge construction activity:

Group Activity

Resources:

Photocopies

Laptops and Classic GeoGebra Software

Photocopies of the history of the Perigal verification and the following construction protocol are given to each group:

Henry Perigal was born in London on 1 April 1801 and died at the age of 97. He created a puzzle known as "The Perigal Puzzle" which is made up of 6 pieces:

J a right-angled triangle,

J a square

J four congruent trapezoids, each with two right angles.

Follow the "Construction Protocol" to assemble the Perigal puzzle:

1. Open a new file in GeoGebra.
2. Hide the coordinate axes by selecting the *Axes* tool from the *View* menu.
3. Activate the **Segment Between Two Points** tool · and construct the segment between points A and B.
4. Activate the **Perpendicular Line** tool · and construct the line *b* that is perpendicular to the segment AB and passes through point A.
5. Activate the **New Point** tool · and construct a point C, on the line b that does not coincide with point A and that the distance to point A is greater than the distance between points A and B.
6. Activate the **Expose/Hide Object** tool to hide the segment AB and the line *b*.
7. Activate the **Polygon** tool · and construct the triangle *ABC*.
8. Activate the **Regular Polygon** tool · and construct three squares, one on each side of the triangle. To do this, mark two vertices of the triangle and indicate that the polygon will have four sides (if the square is inside the triangle, undo with Ctrl-Z and now mark the vertices of the triangle in the opposite order). First build the square on the hypotenuse *BC*, so that the square *BCED*, then on the side *CA* so that the square *CAGF* and finally the square on the side *BA,* so that the square *BAHI.*
9. Activate the **Parallel Line** tool \ and construct a line and name it *j* that is parallel to segment *CE* and passes through point G. Also construct another line and name it line *k* which is parallel to segment *ED* through point F.
10. Activate the tool **Intersection Between Two Objects** \ and construct the point J which is the intersection of the two previous lines *j* and *k*. Calculate also the intersection of the line *j* with the segment *AC* and name it K. The intersection of line *k* with segment *AG* and name it L.
11. Activate the **Pobgono** tool · and build the *CKJF, FJG, GJL* and *ALJK.* For each of them right click on dl and choose the *Configuration...* option, in the *Colour* tab choose a different colour for each of them.
12. With the same **Pobgon** tool build another ABIH pobgon *and* change its colour.

13. Activate the slider tool $\overset{a=2}{\longrightarrow}$. and build a slider called **translation**, this should be set from 0 to 1 with an increment of 0.01.

14. Type in the input line: (a) M = J + translation * (C - J) (b) N = J + translation * (B - J) (c) O = J + translation * (D - J) (d) P = J + translation * (E - J)

15. Activate the **Compass** tool and construct the *circle,* with radius given by the distance between points F and J and whose centre is D (mark the three points in that order). (Make sure that you are actually marking point D as the centre of the circle, for this it is suggested to set the translation slider to 0.5).

16. Activate the **Intersection Between Two Objects** tool and construct the point Q that
is the intersection of the *circumferenciap* with the segment *DE.*

17. Type in the input line: R = B + translation * (Q - B)

18. Activate the **Vector Between Two Points** tool · and construct the vector *u* from point J to point M, the vector *v* from point J to point N, the vector *w* from point J to point O, the vector *zi* from point J to point P and the vector *m* from point B to point R.

19. Activate the **Move Object by Vector** tool · and mark in order the *AKJL* pobgon (mark the pobgon in the centre, not the points) and the *u-vector*, then the *FCKJ* pobgon with respect to the *v-vector*, the *FJG* pobgon with respect to the *w-vector*, the *GJL* pobgon with respect to the *zi-VECTOR* and the *ABIH* pobgon with respect to the *m-vector.*

20. Activate the Pick and Move tool and move the **translation** slider to see the effect.

21. Activate the **Expose / Hide Object** tool to hide all objects that are no longer needed: vectors, non-moving pobgons, etc. Activate the

Expose / Hide Label tool AA. to hide all labels that do not do
falia

22. It is possible to activate the animation by right clicking on the slider and in the pop-up menu activate *Automatic Animation*. This starts the animation; on the screen, in the lower left corner, a small button will appear, press it to stop the animation, and press it again to start it again.

23. Modify the size, colour and styles of your construction.

24. Save the file.

After verification with GeoGebra, the statement of the Pythagorean Theorem is formalised on the basis of the students' observations and conclusions.

> *In any right triangle, the area of the square whose side is the hypotenuse is equal to the sum of the areas of the squares whose sides are each of the legs.*

a plication activity:
Individual Activity
Resources:
Photocopy
Sheet and pencil
Calculator (if required)

The photocopy shown below is handed out and the following instructions are given:

The Hindus called the rectangular triangle with sides 5, 12 and 13 the "Indian triangle". They classified the triangles according to whether the difference between the hypotenuse and the greater leg is equal to 1, 2 or 3.

Give an example of a tern whose difference between the hypotenuse and the greater leg of:

1

2

3

Implementation activity:

Individual Activity

Resources:

Photocopy

Sheet and pencil

The photocopy shown below, which is an adaptation of the activity proposed in Alsina et al. (1989, p. 54), is distributed and the following instructions are given:

Bhaskara (1114-1185) was a Hindu mathematician and astronomer, known for creating the formula for solving quadratic equations. And he performed a proof of the Pythagorean theorem, which we will discuss together:

Given the right triangle abc:

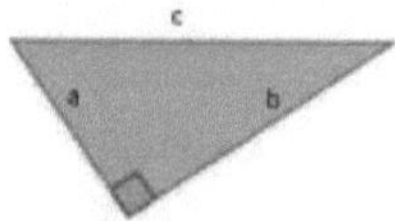

A square is constructed on the hypotenuse, which is divided into four triangles congruent to the given one and a square of side equal to the difference of the legs, i.e. (b - a).

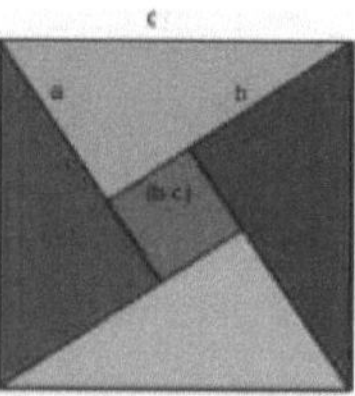

In the following figure it can be seen that the pieces were rearranged, and the figure that turns out to be   the juxtaposition of the squares on the legs was formed:

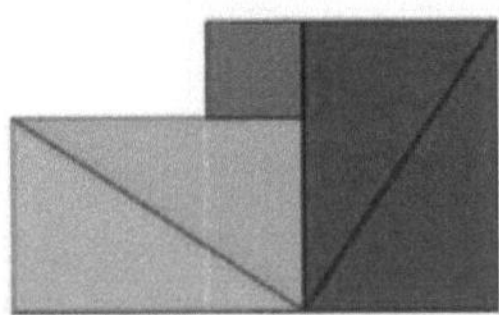

In a geometric way, Bhaskara proved the Pythagorean theorem. It can be translated into algebraic terms by expressing the equality of the drawn figures:

$$c^2 = 4 \cdot \left[\left(\tfrac{1}{2}\right)ab\right] + (b-a)^2 \ (1)$$

$$c^2 = 2ab + b^2 - 2ab + a^2 \ (2)$$

$$c^2 = b^2 + a^2 \ (3)$$

In colloquial language, step (1) of the above test can be described as follows:
The area of the square of side c is equal to the area of the 4 triangles of legs a, b plus the area of the square of side (b-c).

Whereas step (3) can be written, in colloquial language, as follows:
The area of the square of side c is equal to the sum of the areas of the square of side a and the area of the square of side b.

Following the example of Bhaskara's test, complete the following language translation table:

Geometric language	Colloquial language	Algebraic language
		$(m+n)^2 = 4\dfrac{mn}{2} + s^2$
	"The area of the square of side (m+n) is equal to the area of the square of side m, plus the area of the square of side n, plus the area of the two triangles of sides m and n".	$(m+n)^2 = m^2 + n^2 + 2\left(\dfrac{m.n}{2} + \dfrac{m.n}{2}\right)$

- Evaluation Rubric

The rubric was developed on the basis of section 4.1. and the Genially tool was used, which has a free version with no trial period.

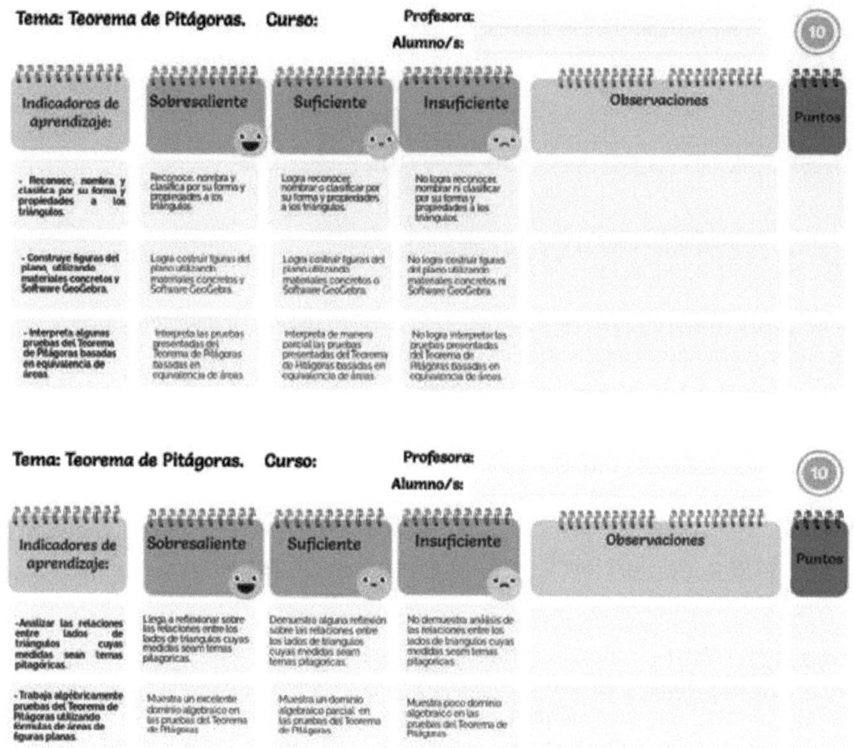

Rubric available at https://view.genial.ly/5f73fa26bd0eb20d80bb7b76/horizontal-infographic-review-rubrica-tesina-secuencias-copia

Chapter 5

Conclusions

Not to forget the concrete origin of mathematics and the historical processes of its evolution.
P.Puig Adam (1951)

5.1 Conclusions

During this work, the importance of the teaching of Geometry at the intermediate level, and above all of the Pythagorean Theorem, and how the proofs made throughout history contribute to its teaching, was highlighted. The main emphasis was on access to the historical construction of knowledge, understanding that this is an inexhaustible source of didactic material, which can be used in didactic sequences in the introduction of the content, with historical notes, used in the development of the sequences to solve some activities, to solve problems, indicating to which mathematician it belongs, to develop capacities proper to the mathematical task, such as reasoning, communication and problem solving.

In Chapter 2, the terms demonstration, proof and explanation were distinguished, and a taxonomy of proofs was made, differentiating the Pragmatic proofs (of verification, generic model and practice), from the Intellectual ones, these in turn into those of visual support and those of formal demonstration (algebraic, quaternionic, geometric and dynamic).

This classification made it possible to catalogue the proofs of the Pythagorean Theorem carried out throughout history, which were developed in Chapter 3. It shows that, as the Florencia Veronica Aspera Matematica evolved, the proofs of the Pythagorean Theorem evolved along with it. In its beginnings they were pragmatic of gendric model, then they were intellectual, some of visual support and others of formal and geometric demonstration. Finally, with the appearance of algebra in the contemporary age, formal, geometric and algebraic proofs emerged. Without prejudice to the fact that the typology of proofs used in some stages is used again in others.

In addition, in this work a historical review of the proofs of the Pythagorean Theorem was carried out, creating a reservoir so that teachers can resort to it whenever they wish. Two didactic sequences were developed that use these proofs in knowledge construction and application activities, both individual and group, from a constructivist paradigm and with an investigative approach. In order to carry them out, we took into account the role of demonstration in the mathematics class, the taxonomy carried out to approach geometric notions and the functions of demonstration. All this under the current legislation in Argentina and taking into account the incorporation of ICT and mathematical potential in the activities of the sequences.

Indeed, the History of Mathematics enriches our teaching and learning practices, it is our challenge to include it in our classes as proposed in this research work.

Questionnaire Prof. Aspera Florencia

All information you provide will be kept strictly confidential, and your name will not appear in any report of the results of this study. Your participation is voluntary and you do not have to answer any questions you do not wish to answer. The purpose of this questionnaire is to collect information about the Pythagorean Theorem and its teaching and learning methods. Your answers are very important because it will allow me to do my thesis for my degree in Mathematics at **the** University Juan Agustin **Maza**. **Thank you very much** for your time, it won't take **more than** 5 minutes. *Compulsory

1. E-mail address *

2. Do you agree to participate in this questionnaire? *

Mark only one oval.

YES

NO

3. First name and Surname *

4. Age * Age * Age * Age * Age * Age * Age * Age * Age * Age * Age * Age

5. Sex * Sex * Sex * Sex * Sex * Sex * Sex * Sex * Sex * Sex * Sex * Sex

Mark only one oval.

FEMALE

MALE

6. Occupation

Mark only one oval.

Student *Skip to question 17*

Mathematics Teacher *Skip to question 7*

Other:

Teacher

The following questions are intended for teachers of mathematics.

7. If you are a teacher, you are o practising o in: *

Mark only one oval.

    EEscuelaPublica
    ESchool   Private
    Both

8. Do you remember in which year/s the Pythagorean Theorem is taught? *

Mark only onevalo.

1°

2°

3°

4°

5°

6°

I don't remember

9. Could you state the Pythagorean Theorem, if so, please do so. *

10. When you teach it/do you include any demonstrations in your teaching practices?

Mark only one oval.

YES

NO

MAYBE

11. What kind of Tests or Demonstrations do you perform? *

Mark only onevalo.

Algebraic proofs: based on relations between sides and segments
Geometric tests: based on comparisons of areas.
Dynamic tests: based on the concepts of mass, speed, strength, etc.
Quaternionic tests: based on vector operations
Other:

12. If you answered yes to the previous question, which one? *

13. At what point in your internship do you include a demonstration?

Mark only one oval.

Prior to giving the Theorem
Immediately after
enunciate it as a
exercise more

14. What types of exercises do you use in your classes? *

Mark only one oval.

Find one side known the other two
In a problematic situation
To check if the given triangle is rectangular
Other:

15. Do you use this Theorem in any subsequent content? *

Mark only one oval.

YES

NO

MAYBE

16. Do you consider the teaching-learning of the Pythagorean Theorem to be relevant?
Why? *

Mathematics students and non-teachers

The following questions are intended for students o not teachers of mathematics.

17. Studying:

Mark only one oval.

Public School
Private School

18. Do you remember in which year/s of high school you were taught the Pythagorean
Theorem? *

Mark only one oval.

1°

2°

3°

4°

5°

6°

I don't remember

19. Could you state the Pythagorean Theorem, if so, please do so. *

20. What types of exercises do you use in your classes? *

Mark only one oval.

Find one side known the other two
In a problematic situation
To check if the given triangle is rectangular
Other:

21. When you were taught it, did you include any demonstration or verification of it? *

Mark only one oval.

YES
NO
I DO NOT REMEMBER

References Bibliographic references

Acevedo, P. H. (2011). *El Teorema de Pitagoras construction de algunos recursos didacticos* (Master's Thesis). Bogota, Cundinamarca, Colombia: Universidad Nacional de Colombia Sede Bogota.

Alsina, C. (2010). *The sect of numbers*. Barcelona: RBA.

Alsina, C., Burgues and Fortuny, J. M. (1987). *Invitation to the didactics of geometry*. Madrid: Srntesis.

Arsac, G. (1987). The origin of demonstration: an essay on didactic epistemology. In *Recherches en Didactique des Mathematiques*, Vol 8, n° 3 (pp. 267-312).

Aznar, E. (2007). Pythagoras, Greek mathematician and philosopher. Retrieved September 9, 2020, from https://www.ugr.es/~eaznar/pitagoras.htm

Balacheff, N. (1987). Processus de preuve et situations de validation. *Educational Studies in Mathematics,* 18, pp. 147-176.

Balacheff, N. (1988). *Le contrat et la coutume. Deux registres des interactions didactiques*. In C. Laborde (Ed.), Actes du Premier Colloques Franco-Allemand de Didactique des Mathematiques et de l'informatique (pp. 15-26). Grenoble: La Pensee Sauvage.

Balacheff, N. (2000). *Los procesos de prueba en los alumnos de matematicas*. Bogota: A Teaching Company. Universidad de los Andes.

Baroody, A. and Dowker, A. (Eds.) (2003*). Studies of mathematical thinking and learning. The development of arithmetic concepts and skills: construction of adaptive experience*. Lawrence Erlbaum Associates Publishers.

Boyer, C.B. (1986). *Historia de lasMatematicas*. Alianza Universidad. Madrid.

Brousseau, G. (1986). Fundamentos y metodos de la didactica, RDM N° 9 (3). Spanish version published by *Facultad de Matematica, Astronomia y Fisica de la Universidad de Cordoba*.

Burton, D. (2011). *The History of Mathematics an introduction* (7th ed., p. 34). New York: McGraw-Hill.

Chevallard, Y. (1985). *La transposition didactique; du savoir savant au savoir enseigne*, Paris, La Pensee Sauvage.

Crespo, C. (2005). *The role of mathematical argumentation in school discourse. The strategy of deduction by reduction to the absurd*. Unpublished Master's thesis. CICATA-IPN, Mexico.

Crespo, C., Farfan, R. and Lezama, J. (2010). Arguments and demonstrations: a view of the influence of sociocultural settings. *Revista Latinoamericana de Investigation en Matematica Educativa*, vol. 13, num. 3, November, 2010, pp. 283-306 Comite Latinoamericano de Matematica Educativa Distrito Federal, Organismo International.

Damisa, C. and Ponzetti S. (2015). *The mathematical object and its representations: ^What you see is what it is?* Curem5.

De Castro, C. (2007). La evaluation de metodos para la ensenanza y el aprendizaje de las matematicas en la Educación Infantil. Union. *Revista Iberoamericana de Education Matematica*, 11, 59-77.

De Villiers, M. (1993). *The role of paper and the role of demonstration in mathematics*. Epsilon, *26*, 15-30. Original 1990.

Direction General de Cultura y Education de la provincia de Buenos Aires (2010). *Diseno curricular para la Education Secundaria Ciclo Superior ES4: Matematica.* coordinated by Claudia Bracchi -1a ed.- La Plata.

Direction General de Cultura y Education (2007). *Diseno Curricular para la Educación Secundaria 2° ano (SB),* Gobierno de la Provincia de Buenos Aires, Buenos Aires.

Dreyfus, T. (1999). Why Johnny can't prove? *Educational Studies in Mathematics*, 38(1/3), 85-109.

Dunham,W. (1995). *The Universe of Mathematics*. Piramide. Madrid. Chap.H.

Duval, R. (1993). Registres de representations semiotique et fonctionnement cognitif de la pensee, *Annales de Didactique et de Sciences Cognitives, irem de Strasbourg*, France, 5, 37-65. Retrieved from: https://mathinfo.unistra.fr/fileadmin/upload/IREM/Publications/Annales_didactique/vol_0 5/adsc5_1993-003.pdf

Duval, R. (1999). *Argumentar, demostrar, explicar: ^continuidado ruptura cognitiva?* Mexico: Grupo Editorial Iberoamerica.

Duval, R. (2007). *Cognitive functioning and the understanding of mathematical processes of proof.* In P. Boero (Ed.), Theorems in schools: from history and cognition to classroom practice, (pp. 137-161). Rotterdam, Netherland: Sense Publishers.

Euclid (1991). *Elements. Libros I-IV.* Madrid: Gredos.

Garda Pena, S. and Lopez Escudero, O. (2008). La ensenanza de la Geometria. Mexico, D.F. Second edition, 2011, 174 pp. *Education Matematica*, vol. 24, num. 2, August, 2012, pp. 135-140 Grupo Santillana Mexico Distrito Federal, Mexico.

Godino, J. D. and Recio, A. M. (2001). Institutional meanings of demonstration. Implications for mathematics education. *Revista Ensenanza de las ciencias*, 19 (3), (pp. 405-414).

Gonzalez Urbaneja, P. (2001). *Pitagoras, el Filosofo del numero.* Nivola, Madrid.

Gonzalez Urbaneja, P. (2004). La Historia de la Matematica como recurso didactico e instrumento para enriquecer culturalmente su ensenanza. *Revista SUMA (F'ESPM).* num.45, pp.17-28.

Gonzalez Urbaneja, P. (2008). *The so-called Pythagoras theorem. A geometric history of 4,000 years.* SIGMA.

Guillen, G. (1997). *Van Hiele's model applied to the geometry of solids. Observation of learning processes* (Doctoral thesis). Valencia: Universitat de Valencia.

Hanna, G. and Jahnke, H. (1996). Proof and proving. In A. Bishop, K. Clements, C. Keitel, J. Kilpatrick and C. Laborde (Eds.). Laborde (Eds.), *International Handbook of Mathematics Education* (pp. 877-908). Dordrecht: Kluwer Academic Publishers.

Harel, G. and Sowder, L. (1998). Student's Proof Schemes: Results from exploratory studies. In E. Dubinsky, A. Schoenfeld & J. Kaput (Eds.), *Research on Collegiate Mathematics Education* (Vol. 3, pp. 234-283). Providence, RI: American Mathematical Society.

Hernandez Cruz, C. (2019). *El teorema de Pitagoras, pretexto y contexto para la ensenanza de la Geometria* (Magister en Ensenanza de las Ciencias Exactas y Naturales). National University of Colombia, Faculty of Sciences.

Hernandez, E. (2017*). Classification of isometries in the plane using GeoGebra.* Universidad Autonoma de Madrid. *40° Reunion de Educacion Matematica Union Matematica Argentina.*

Hilbert, D. (1898/1899b). *Elemente der Euklidischen Geometrie,* (MS. Vorlesung, WS 1898/9). ausgearbeitet von Hans von Schaper. Niedersachsische Staats- und Universitatsbibliothek Gottingen, Handschriftenabteilung, Cod. D. Hilbert 552. Published in U. Majer & M. Hallett (2004), pp. 302-402.

Ibanes, M. and Ortega, T. (1997). Demonstration in mathematics. Clasificacion y ejemplos en el marco de la educacion secundaria. *In EducacionMatematica.* Vol. 9 n°2. (pp. 65-104) Mexico: Grupo Editorial Iberoamerica.

Jaen Sanchez, M. (2012). *Pitagoras, el teorema de Pitagoras.* Barcelona. RBA.

Kepler, J. (1596). *The Secret of the Universe (*1ª ed.). Madrid: Alianza. Trad., introduction and notes by E. Rada.

Kline, M. (1978). *The failure of modern mathematics.* Siglo XXI, Madrid.

Krantz, S. (2007). *The history and concept of mathematical proof.*

Lamarca, J. M. (2006). A new geometric-algebraic demonstration of the Pythagorean theorem. *Revista SUMA (FESPM),* num.52, p.142.

Lazaro Carreter, F. (1999). Esprntu de Geometria (in EL DARDO EN LA PALABRA). EL PA^S, 5/12/99.

Loomis, E. (1968). The Pythagorean Proposition. Its Demonstrations Analyzed and Classified. *National Council of Teachers of Mathematics (Classics in Mathematics Education).* Washington.

Mansfield, D. and Wildberger, N. (2017). *A Babylonian tablet hides the world's oldest trigonometric table.* Retrieved September 9, 2020, from https://www.agenciasinc.es/Noticias/Una-tablillababilonica-esconde-la-tabla-trigonometrica- most-ancient-in-the-world.

Mieville, D. (1981). Explication et discours didactique de la mathematique. *Revue europeenne des sciences sociales et Cahiers Vilfredo. ParetoXIX,* 56, pp.115- 152.

Ministerio de Cultura y Educacion de la Nacion (1995). *Contenidos Basicos Comunes para la Educación General Basica,* Buenos Aires.

Mishra, P., and Koehler, M. J. (2006). Technological pedagogical content knowledge: A framework for integrating technology in teacher knowledge. *Teachers College Record*, 108(6), 1017-1054.

Principles and Standards for Mathematics Education (2000). Seville: *Sociedad Andaluza de Educacion Matematica Thales*.

Needham, J. (1959). Mathematics and the Sciences of the Heaven and the Earth. *Science and Civilisation in China,* Vol. 3, Cambridge, Cambridge University Press.

Perigal, H. (1872). *On the geometric dissections and transformations*. England: [s.n.].

Polya, G. (1984). *Como plantear y resolver problemas*. Mexico: Trillas.

Prior Martinez, J. and Torregrosa Girones, G. (2013). Configural reasoning and verification procedures in geometric context. *Revista Latinoamericana de Investigation en Matematica Educativa,* 16 (3), 339-368. https://dx.doi.org/10.12802/relime.13.1633

Puig Adam, P. (1951). Decalogo de la Didactica Matematica Media, *Gaceta Matematica,* 1ª serie, tomo 7, n°5-6, Madrid.

Puig, L and Cerdan, F. (1995). *Problemas aritmeticos escolares*. Madrid: Smtesis.

Rodriguez, M. (2012). Problem solving. In M. Pochulu and M. Rodriguez (eds.), *Education matematica. Aportes a la formación docente desde distintos enfoques teoricos*, 155-177. Buenos Aires: ungs-eduvim

Sanchez, C. (2012). History as a didactic resource: the case of Euclid's Elements. *Tecne Episteme YDidaxis: TED*, (32). doi: 10.17227/ted.num32-1860

Santalo, L. (1966). *La matematica en la escuela secundaria*. Buenos Aires: Eudeba.

Santalo, L. (1975). *La Education matematica hoy*. Teide, Barcelona.

Sorando Muzas, J. (2020). Historia resolucion de problemas - Matematicas en tu mundo. Retrieved 14 August2020 , from http://matematicasmundo.ftp.catedu.es/PROBLEMAS/problemas_importancia_historica.htm

Swetz, F., and Katz, V. (2011). Mathematical treasures - Zhoubi suanjing. Retrieved September 9, 2020, from https://www.maa.org/book/export/html/116902.

Torregrosa-Girones, G., Delicado, M., Martinez, M., & Ciscar, S. (2010). Teacher's conceptions of testing and dynamic software. Development in a virtual learning environment. *Journal of education*, 352, 379-404.

Tostado, F. (2013). The Pythagorean theorem before Pythagoras. Retrieved 20 August 2020, from https://franciscojaviertostado.com/2013/08/09/el-teorema-de-pitagoras-antes-de-pitagoras/

Vega, L. (1990). *La trama de la demostracion (Los griegos y la razon tejedora de pruebas)*. Madrid: Alianza.

Verne, J. (1865). *From the Earth to the Moon* (Mauro Armino, Trad.) (4th ed.). Spain: EDAL S.L.

Vitruvius Polion, M. (2009). *The ten books of Architecture*. With prologue by D. Rodriguez, "Vitruvius and the skin of classicism", pp. 11-51. Madrid: Editorial Alianza.

Waldegg, G. (1998). Constructivist principles for mathematics education. *Ema Journal. Investigation E Innovation En Educacion Matematica, (*VOL. 4, N° 1), 16-31.

Printed by Books on Demand GmbH, Norderstedt / Germany